To my teachers, Evgeniy Alekseevich Gorin
and Aleksandr Isaakovich Plotkin

Contents

CHAPTER 1

Introduction

The study of the geometry of convex bodies based on information about sections and projections of these bodies has important applications to many areas of mathematics and science. In this book we discuss a new Fourier analytic approach that has recently been developed. The idea is to express certain geometric properties of bodies in Fourier analytic terms and then use methods of harmonic analysis to solve geometric problems. We start with a few examples.

The Fourier analytic approach to sections of convex bodies is based on certain formulas expressing the volume of sections in terms of the Fourier transform of powers of the Minkowski functional of a body. The first formula of this kind was known to Laplace [La], who wrote that the $(n-1)$-dimensional volume of the central hyperplane section of the unit cube Q_n in $\mathbb{R}^n$ perpendicular to the main diagonal is equal to

$$\text{(1.1)} \qquad \mathrm{Vol}_{n-1}(Q_n \cap (\frac{1}{\sqrt{n}}, ..., \frac{1}{\sqrt{n}})^{\perp}) = \frac{1}{\pi}\int_{-\infty}^{\infty}\left(\frac{\sin(r/\sqrt{n})}{r/\sqrt{n}}\right)^n dr.$$

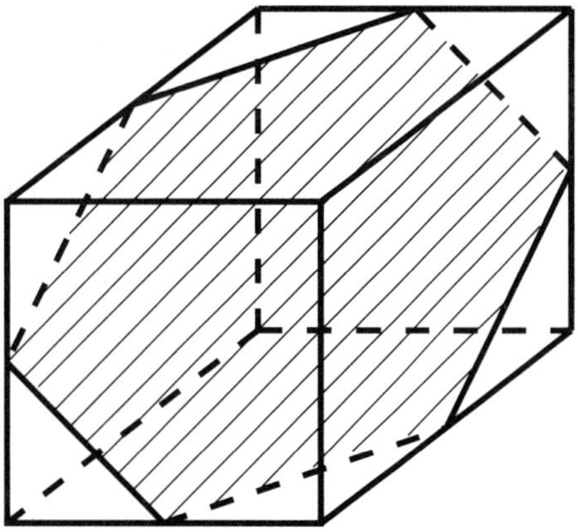

Section orthogonal to $(1, 1, ..., 1)$.

The latter integral is exactly what appears in the central limit theorem for the sums of uniformly distributed random variables, and it is easy to see that the volume of the diagonal section tends to $6/\sqrt{\pi}$, as $n \to \infty$. This result is quite surprising, because it contradicts our intuition that the section perpendicular to the main diagonal has the maximal volume among all central hyperplane sections of the cube. In fact, in every dimension the section perpendicular to the vector $(1, 1, 0, ..., 0)$ has volume $\sqrt{2} > 6/\sqrt{\pi}$.

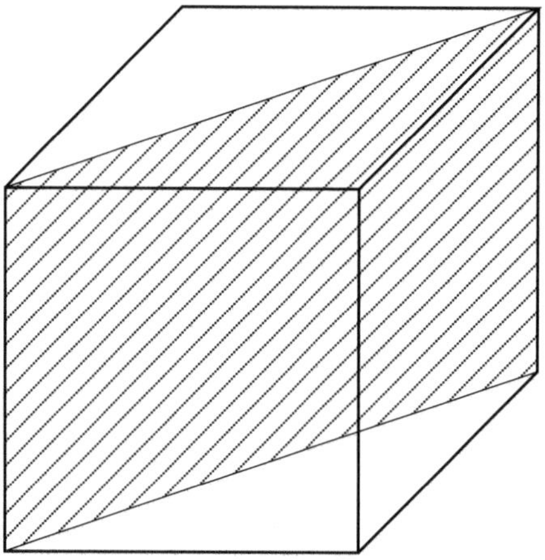

Section orthogonal to $(1, 1, 0, ..., 0)$.

The problem of finding the maximal volume of central hyperplane sections of the cube remained open until the 1980's, when first Hensley [He1] proved that these volumes are bounded from above by a constant not depending on the dimension, and then Ball [Ba1] found that this constant is exactly $\sqrt{2}$. Both Hensley and Ball used a general Fourier transform formula for the volume of hyperplane sections of the cube first mentioned in the literature by Polya [Po1]:

$$\text{Vol}_{n-1}(Q_n \cap \xi^\perp) = \frac{1}{\pi} \int_{-\infty}^{\infty} \prod_{k=1}^{n} \frac{\sin(r\xi_k)}{r\xi_k}\, dr, \tag{1.2}$$

where, for $\xi \in S^{n-1}$, $\xi^\perp = \{x \in \mathbb{R}^n : (x, \xi) = 0\}$ is the central hyperplane perpendicular to the vector ξ. A proof of Ball's result is presented in Chapter 7, Theorem 7.1.

Meyer and Pajor [MeP] found a similar formula for the volume of central hyperplane sections of ℓ_q-balls:

$$\text{Vol}_{n-1}(B_q^n \cap \xi^\perp) = \frac{q}{\pi(n-1)\Gamma((n-1)/q)} \int_0^\infty \prod_{k=1}^{n} \gamma_q(t\xi_k)\, dt, \tag{1.3}$$

where γ_q is the Fourier transform of the function $\exp(-|\cdot|^q)$ on $\mathbb{R}$, and B_q^n is the unit ball of the space ℓ_q^n :

$$B_q^n = \{x \in \mathbb{R}^n : \|x\|_q = \left(\sum_{k=1}^{n} |x_k|^q\right)^{1/q} \le 1\}. \tag{1.4}$$

Our work in the area of convex geometry was initiated by formula (1.3). Having known the expression for the Fourier transform of powers of ℓ_q-norms, Lemma 3.6, one can recognize in the right-hand side of (1.3) the Fourier transform of the function $\|\cdot\|_q^{-n+1}$ at the point ξ, up to a constant not depending on q. This suggests that (1.3) is a particular case of a more general formula, and this is indeed the case, as proved in [K7]:

THEOREM 1.1. *For any origin-symmetric star body K in $\mathbb{R}^n$ and any $\xi \in S^{n-1}$,*

$$\text{Vol}_{n-1}(K \cap \xi^\perp) = \frac{1}{\pi(n-1)} (\|\cdot\|_K^{-n+1})^\wedge(\xi), \tag{1.5}$$

where $\|\cdot\|_K$ *is the Minkowski functional of* K. *The Fourier transform of* $\|\cdot\|_K^{-n+1}$ *is considered in the sense of distributions and as such coincides with a homogeneous of degree* -1 *function on* $\mathbb{R}^n$, *which is continuous on* $\mathbb{R}^n \setminus \{0\}$.

This formula has many geometric applications. For example, the Minkowski uniqueness theorem (see [Ga3, Th. 7.2.3]) states that every origin-symmetric star body is uniquely determined by the volumes of its central hyperplane sections. By (1.5), this result immediately follows from the uniqueness theorem for the Fourier transform, because $(\|\cdot\|_K^{-n+1})^\wedge$ is an even homogeneous function on $\mathbb{R}^n$.

The Fourier analytic approach is quite effective in the study of intersection bodies. Let K and L be origin-symmetric star bodies in $\mathbb{R}^n$. Following Lutwak [Lu1], we say that K is the *intersection body of* L if the radius of K in every direction is equal to the volume of the central hyperplane section of L perpendicular to this direction, i.e. for every $\xi \in S^{n-1}$,

$$\|\xi\|_K^{-1} = \mathrm{Vol}_{n-1}(L \cap \xi^\perp). \tag{1.6}$$

A more general class of *intersection bodies* can be defined as the closure in the radial metric of the class of intersection bodies of star bodies.

Using formula (1.5), one can get a Fourier analytic characterization of intersection bodies. In fact, applying (1.5) to the right-hand side of (1.6), we get

$$\|\xi\|_K^{-1} = \frac{1}{\pi(n-1)}(\|\cdot\|_L^{-n+1})^\wedge(\xi).$$

Since the Fourier transform of even distributions is self-invertible (up to a constant),

$$(\|\cdot\|_K^{-1})^\wedge(\xi) = \frac{(2\pi)^n}{\pi(n-1)}\|\xi\|_L^{-n+1}.$$

In particular, the Fourier transform of $\|\cdot\|_K^{-1}$ is non-negative. This consideration leads to a Fourier analytic characterization of intersection bodies that was proved in [K10]:

THEOREM 1.2. *An origin-symmetric star body* K *in* $\mathbb{R}^n$ *is an intersection body if and only if the Fourier transform of* $\|\cdot\|_K^{-1}$ *is a positive distribution, i.e. its action on any non-negative test function gives a non-negative result.*

In Section 4.1. we prove Theorem 1.2. This characterization allows us to give many examples of bodies that are and are not intersection bodies. For instance, as proved in [K10], the unit ball of any finite dimensional subspace of L_q with $0 < q \le 2$ is an intersection body. On the other hand, as proved in [K8], the unit balls of the spaces ℓ_q^n, $2 < q < \infty$, are intersection bodies if the dimension $n \le 4$, but they are not intersection bodies if $n \ge 5$. We prove these facts in Section 4.3, Theorems 4.11 and 4.13.

Intersection bodies played an important role in the solution to the Busemann-Petty problem. Posed in 1956 in [BP], the Busemann-Petty problem asks the following question. Let K and L be origin-symmetric convex bodies in $\mathbb{R}^n$ so that the $(n-1)$-dimensional volume of every central hyperplane section of K is smaller than the same for L. Does it follow that the n-dimensional volume of K is smaller than the n-dimensional volume of L?

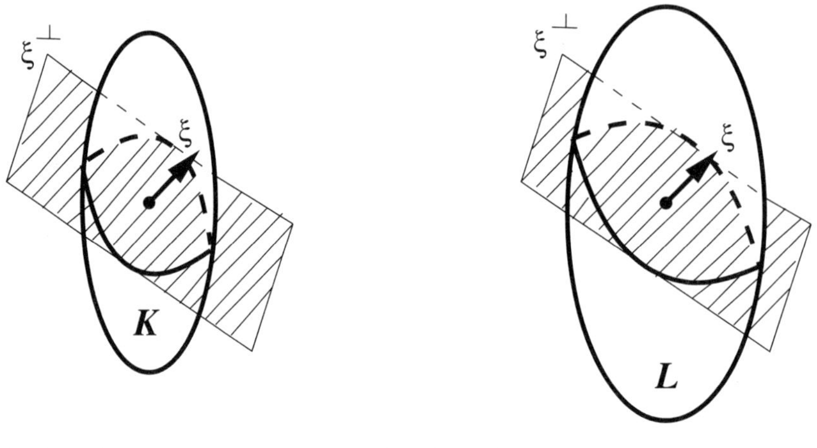

The Cavalieri principle and the Minkowski uniqueness theorem mentioned above provide strong intuition in the direction of the affirmative answer. Indeed, how can the answer to the Busemann-Petty problem be negative if every origin-symmetric convex body is uniquely determined by the volumes of its central hyperplane sections? Another piece of evidence in favor of the affirmative answer is that it is true in dimension $n = 2$, because the assumption of the problem implies that $K \subset L$.

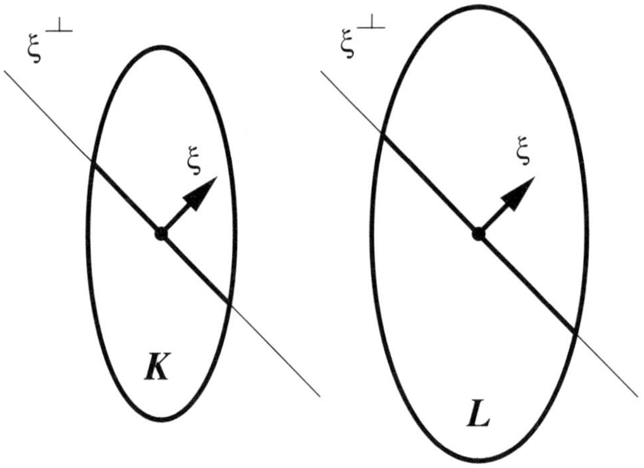

Also, during the first twenty years the results related to the problem all went in the positive direction. Busemann [Bu2] proved that the answer is affirmative if K is an ellipsoid. Hadwiger [Ha2] gave an affirmative answer in the case where K and L are solids of revolution in $\mathbb{R}^3$.

It came as a big surprise when, in 1975, Larman and Rogers [LR] used probabilistic methods to prove that the answer is negative if the dimension $n \geq 12$. Ball [Ba2] used his $\sqrt{2}$ bound mentioned above to construct counterexamples for $n \geq 10$. Ball's examples are the most "visible" so far, if one can say that about objects in $\mathbb{R}^{10}$: the body with smaller sections is the unit cube and the body with larger sections is the Euclidean ball whose radius is chosen in such a way that the volume of each central hyperplane section is equal to $\sqrt{2}$. Giannopoulos [Gi] and Bourgain [Bo1] constructed counterexamples for $n \geq 7$. Papadimitrakis [Pa] proved that the answer is negative for $n \geq 5$, and Gardner [Ga1] and Zhang [Zh2] proved the same by constructing the first examples of non-intersection bodies in $\mathbb{R}^5$. They used Lutwak's connection with intersection bodies (see Theorem 1.3). In another surprising development, Gardner [Ga2] proved that every origin-symmetric convex body in $\mathbb{R}^3$ is an intersection body, and, hence, the answer to the Busemann-Petty problem in dimension 3 is affirmative. For several years the problem was considered as completely solved, since the paper [Zh2] published in 1994 claimed that the unit cube in $\mathbb{R}^4$ was not an intersection body, which would imply a negative answer in dimension 4. However, it was proved in [K8] in 1997 that the cube in $\mathbb{R}^4$ is an intersection body, which reopened the four-dimensional case and suggested that the answer for $n = 4$ must be affirmative instead of negative. After this, Zhang [Zh4] proved that the answer to the Busemann-Petty problem in dimension 4 is indeed affirmative, and a unified Fourier analytic solution to the problem in all dimensions was given in [GKS].

The last steps of the solution were made using two different approaches. Both of them used Lutwak's concept of an intersection body. However, the study of intersection bodies, which requires inversion of the spherical Radon transform in a special situation, was done differently. The first, more geometric, approach was developed by Gardner [Ga1], [Ga2] and Zhang [Zh1], [Zh4]. The second, Fourier analytic, approach of [K10] and [GKS] provided a unified solution to the problem in all dimensions at the same time. This approach clearly exposed the reason behind the transition between dimensions 4 and 5 and traced this transition to the fact that convexity controls the derivatives of the second order, but not the derivatives of higher orders. The main ideas of the geometric approach can be found in [Ga3], so we concentrate on the Fourier analytic part.

The Fourier analytic proof in [GKS] is based on three main ingredients. The first is a connection between intersection bodies and the Busemann-Petty problem found by Lutwak [Lu1] and slightly modified by Gardner [Ga1] and Zhang [Zh1], [Zh2]. First, if K is an intersection body, then the answer to the question of the Busemann-Petty problem is affirmative for this K and any origin-symmetric star body L. On the other hand, if L is an origin-symmetric convex body that is not an intersection body, then one can perturb L to construct a counterexample to the Busemann-Petty problem. Using these connections, Lutwak reduced the Busemann-Petty problem to the study of intersection bodies:

THEOREM 1.3. *A solution to the Busemann-Petty problem in $\mathbb{R}^n$ is affirmative if and only if every origin-symmetric convex body in $\mathbb{R}^n$ is an intersection body.*

We prove Lutwak's connections twice. First we present the original geometric proofs in Section 5.1, and then we prove more general facts in Section 5.2 using Fourier methods.

The solution in [GKS] also uses the Fourier analytic characterization of intersection bodies from Theorem 1.2 and the following formula proved in [GKS]:

THEOREM 1.4. *Let D be an infinitely smooth origin-symmetric convex body in $\mathbb{R}^n$, $k \in \mathbb{N} \cup \{0\}$, $k \neq n-1$, $\xi \in S^{n-1}$. Then if k is an even integer,*

$$(\|\cdot\|_D^{-n+k+1})^\wedge(\xi) = (-1)^{k/2}\pi(n-k-1)A_{D,\xi}^{(k)}(0).$$

If k is an odd integer,

$$(\|\cdot\|_D^{-n+k+1})^\wedge(\xi)$$
$$= (-1)^{(k+1)/2}2(n-k-1)k!\int_0^\infty \frac{A_{D,\xi}(z) - A_{D,\xi}(0) - \dots - A_{D,\xi}^{(k-1)}(0)\frac{z^{k-1}}{(k-1)!}}{z^{k+1}}\,dz.$$

Here $A^{(k)}_{D,\xi}(0)$ *is the derivative of the order* k *at zero of the parallel section function* $A_{D,\xi}$ *of the body* D *in the direction of* ξ *defined by*

$$A_{D,\xi}(z) = \mathrm{Vol}_{n-1}(D \cap \{\xi^\perp + z\xi\}), \quad z \in \mathbb{R}.$$

In both cases, the Fourier transform of $\|\cdot\|_D^{-n+k+1}$ *is considered in the sense of distributions and coincides with a homogeneous of degree* $-k-1$ *function on* $\mathbb{R}^n$, *which is infinitely smooth on* $\mathbb{R}^n \setminus \{0\}$.

Let us show how the solution to the Busemann-Petty problem follows from these results. By a simple approximation argument, it is enough to consider the case where the bodies K and L are infinitely smooth. Suppose that the dimension $n=4$. Putting $n=4,\ k=2$ in Theorem 1.4, we get that, for every $\xi \in S^{n-1}$,

$$(\|\cdot\|_K^{-1})^\wedge(\xi) = -\pi A''_{K,\xi}(0).$$

Now we use the fact that the body K is origin-symmetric convex. By a well-known theorem of Brunn, Theorem 2.3, the central section has maximal volume among all hyperplane sections of K perpendicular to a given direction. This means that the function $A_{K,\xi}$ has maximum at zero, so the second derivative of this function at zero is less than or equal to zero. Hence, the Fourier transform of $\|\cdot\|_K^{-1}$ is non-negative for every origin-symmetric convex body K in $\mathbb{R}^4$. By the Fourier characterization of intersection bodies, Theorem 1.2, every origin-symmetric convex body in $\mathbb{R}^4$ is an intersection body, and the affirmative solution to the Busemann-Petty problem follows from Lutwak's connection, Theorem 1.3.

On the other hand, suppose that the dimension $n=5$. Put $k=3$ in Theorem 1.4. We use the second formula of this theorem, and we are essentially looking at the third derivative of the function $A_{K,\xi}$. But convexity does not control the third derivatives! With this in mind, it is quite easy to construct an example of an infinitely smooth convex body for which the Fourier transform of $\|\cdot\|_K^{-1}$ is sign-changing (we construct such an example in Corollary 4.4). Again by the Fourier analytic characterization of intersection bodies, this gives an example of an infinitely smooth convex body in $\mathbb{R}^5$ that is not an intersection body, and a negative solution to the Busemann-Petty problem in dimension 5 follows from Lutwak's connection, Theorem 1.3.

This solution clearly shows that the transition between the dimensions 4 and 5 in the Busemann-Petty problem occurs due to the fact that convexity controls the second derivatives, but not the derivatives of higher orders.

We finish the preview by showing a Fourier analytic proof of the first Lutwak connection. This proof is short and gives a good idea of how the Fourier approach works. We are going to prove a more general result later in Theorem 5.5.

THEOREM 1.5. *Let* K *be an infinitely smooth origin-symmetric intersection body in* $\mathbb{R}^n$, *and let* L *be any origin-symmetric star body in* $\mathbb{R}^n$ *so that, for every* $\xi \in S^{n-1}$,

$$\mathrm{Vol}_{n-1}(K \cap \xi^\perp) \le \mathrm{Vol}_{n-1}(L \cap \xi^\perp). \tag{1.7}$$

Then $\mathrm{Vol}_n(K) \le \mathrm{Vol}_n(L)$.

PROOF. By Theorem 1.1, the Fourier transforms of $\|\cdot\|_K^{-n+1}$ and $\|\cdot\|_L^{-n+1}$ are continuous functions on $\mathbb{R}^n \setminus \{0\}$, and condition (1.7) can be written in the form

$$(\|\cdot\|_K^{-n+1})^\wedge(\xi) \le (\|\cdot\|_L^{-n+1})^\wedge(\xi), \quad \forall \xi \in S^{n-1}. \tag{1.8}$$

Since K is infinitely smooth, by Corollary 3.17, the Fourier transform of $\|\cdot\|_K^{-1}$ is a continuous function on the sphere. Also K is an intersection body, so, by Theorem 1.2, this function is non-negative. We have

$$\int_{S^{n-1}} (\|\cdot\|_K^{-1})^\wedge(\xi)(\|\cdot\|_K^{-n+1})^\wedge(\xi)\, d\xi \le \int_{S^{n-1}} (\|\cdot\|_K^{-1})^\wedge(\xi)(\|\cdot\|_L^{-n+1})^\wedge(\xi)\, d\xi.$$

Now we can use a version of Parseval's formula on the sphere, Lemma 3.22 to remove the Fourier transforms:

$$\int_{S^{n-1}} \|x\|_K^{-1}\|x\|_K^{-n+1}\, dx \le \int_{S^{n-1}} \|x\|_K^{-1}\|x\|_L^{-n+1}\, dx.$$

By the polar formula for the volume (2.4), the left-hand side of the latter inequality is equal to n times the n-dimensional volume of K. Using Hölder's inequality to estimate the right-hand side, we get

$$n\mathrm{Vol}_n(K) \le \left(\int_{S^{n-1}} \|x\|_K^{-n}\, dx\right)^{1/n} \left(\int_{S^{n-1}} \|x\|_L^{-n}\, dx\right)^{(n-1)/n}$$
$$= n\left(\mathrm{Vol}_n(K)\right)^{1/n}\left(\mathrm{Vol}_n(L)\right)^{(n-1)/n},$$

which implies that the n-dimensional volume of K is smaller than that of L. □

The book is organized in the following way. In Chapter 2, we gather definitions, notation and technical tools needed in the rest of the text. We include the proofs that are difficult to find and do not prove the facts that can easily be found in one of our main references. We refer the reader to the books by Schneider [Sch3] and Gardner [Ga3] for results in convex geometry and to the books by Gelfand and Shilov [GS], Gelfand and Vilenkin [GV] and Rudin [Ru] for information about distributions.

In Chapter 3, we prove several formulas expressing the volume of sections in terms of the Fourier transform, including Theorems 1.1 and 1.4. In Section 3.4, we prove the spherical Parseval formula that was used in the proof of Theorem 1.5 and will be used many more times in the rest of the book.

Intersection bodies are studied in Chapter 4. We start with the Fourier analytic characterization of intersection bodies, Theorem 1.2, in Section 4.1. In the same section, we include a simple geometric characterization of intersection bodies. In particular, if n is an even integer, then a smooth origin-symmetric convex body in $\mathbb{R}^n$ is an intersection body if and only if for every $\xi \in S^{n-1}$,

$$(-1)^{n/2}A_{K,\xi}^{(n-2)}(0) \ge 0.$$

This fact conveniently shows once again that in dimension 4 every such K is an intersection body, because the second derivative is negative. For odd n the formula is slightly more complicated, but also works well for constructing non-intersection bodies in dimension 5.

The concept of a k-intersection body is considered in Section 4.2. These bodies are useful for generalizations of the Busemann-Petty problem and also provide an interesting link between geometry and functional analysis. Let $1 \le k < n$, and let D and L be origin-symmetric star bodies in $\mathbb{R}^n$. We say that D is a *k-intersection body of L* if for every $(n-k)$-dimensional subspace H of $\mathbb{R}^n$

$$\mathrm{Vol}_k(D \cap H^\perp) = \mathrm{Vol}_{n-k}(L \cap H).$$

A more general class of k-intersection bodies can be characterized by the condition that $\|\cdot\|_D^{-k}$ is a positive definite distribution, Theorem 4.8. An important consequence is that every origin-symmetric convex body in $\mathbb{R}^n$ is a k-intersection body for each $k = n-3, n-2, n-1$ such that $k > 0$ (see Corollary 4.9). Note that the original intersection bodies are 1-intersection bodies in this more general setting.

In Section 4.3 we give several examples of bodies that are and that are not k-intersection bodies. We prove that the unit ball of any finite dimensional subspace of L_q, $q \leq 2$, is a k-intersection body for every k. We then prove that the unit balls of the spaces ℓ_q^n, $2 < q \leq \infty$, are k-intersection bodies if and only if $k = n-3, n-2, n-1$ so that $k > 0$, and, therefore, these bodies represent the "extreme" case. The techniques for calculating the Fourier transform used in this section may be of independent interest. In Section 4.4 we present other "extreme" examples by proving the so-called second derivative test for k-intersection bodies, Theorem 4.19. Roughly speaking, if there exists a direction so that the second directional derivatives of the norm in $\mathbb{R}^n$ in this direction vanish in a special way, then the corresponding body is a k-intersection body only for $k \geq n-3$. The examples include q-sums of normed spaces with $q > 2$ and Orlicz spaces with Orlicz functions M satisfying $M'(0) = M''(0) = 0$.

Chapter 5 is devoted to the Busemann-Petty problem. In Section 5.1, we present in detail the unified analytic solution to the problem from [GKS]. In Section 5.2, we consider a generalization of the Busemann-Petty problem. Since the solution to the problem is negative in most dimensions, it is natural to ask for a condition on the behavior of parallel section functions at the origin, which allows us to compare the n-dimensional volumes in all dimensions. We give such a condition in terms of the $(n-4)$-th derivatives of the parallel section function at zero and show that the order $n-4$ is optimal. This result, Theorem 5.7, generalizes the solution to the original Busemann-Petty problem. The proof is purely Fourier analytic.

In Section 5.3, we give a short proof of the affirmative part of the solution to the Busemann-Petty problem (and also the affirmative part of Theorem 5.7) using spherical harmonics.

A very recent result of Zvavitch [Zv2] gives a generalization of the Busemann-Petty problem to arbitrary measures in place of the volume. We present this result in Section 5.4. Let f be a non-negative locally integrable function on $\mathbb{R}^n$, which is also locally integrable on each central hyperplane $\xi^\perp$ in $\mathbb{R}^n$. Let μ be the measure on $\mathbb{R}^n$ with density f. For every closed bounded set $B \in \xi^\perp$ define

$$\mu(B) = \int_B f(x)dx.$$

Given two convex origin-symmetric bodies K and L in $\mathbb{R}^n$ such that, for every $\xi \in S^{n-1}$,

$$\mu(K \cap \xi^\perp) \leq \mu(L \cap \xi^\perp),$$

does it follow that

$$\mu(K) \leq \mu(L)?$$

Surprisingly, for a very large class of functions f, the answer is the same as for the original problem – affirmative if $n \leq 4$ and negative if $n \geq 5$. The proof uses a generalization of (1.5), the spherical Parseval formula from Section 3.4, a clever elementary argument replacing Hölder's inequality in the end of the proof and results on intersection bodies from Chapter 4.

In Section 5.5 we discuss several open problems, in particular, the lower dimensional Busemann-Petty problem asking the same question with k-dimensional sections, $2 \le k < n-2$, in place of hyperplane sections. Bourgain and Zhang [BZ] proved that the answer is negative if $k > 3$. We present a proof of this result from [K14]. The problem remains open for two- and three-dimensional sections, $k = 2, 3$.

In Chapter 6, we show a connection between intersection bodies and the theory of L_p-spaces. Let us discuss this connection in a little more detail. It is a well-known fact going back to P. Levy (see Lemma 6.4) that an n-dimensional space $(\mathbb{R}^n, \|\cdot\|)$ embeds in L_p, $p > 0$, if and only if there exists a finite Borel measure μ on the sphere S^{n-1} in $\mathbb{R}^n$ so that, for every $x \in \mathbb{R}^n$,

$$\|x\|^p = \int_{S^{n-1}} |(x,\xi)|^p d\mu(\xi). \tag{1.9}$$

On the other hand, one can express the volume of hyperplane sections of an origin-symmetric star body in the form

$$\mathrm{Vol}_{n-1}(L \cap \xi^\perp) = \frac{1}{2} \lim_{\epsilon \to 0} \epsilon \int_L |(x,\xi)|^{-1+\epsilon} d\xi.$$

In fact, use the Fubini theorem to write the right-hand side as

$$\lim_{\epsilon \to 0} \epsilon \int_0^\infty t^{-1+\epsilon} A_{L,\xi}(t)\, dt,$$

and it is easy to see that the latter limit equals $A_{L,\xi}(0) = \mathrm{Vol}_{n-1}(L \cap \xi^\perp)$. Thus, a body K is the intersection body of L if and only if, for every $x \in \mathbb{R}^n$,

$$\|x\|_K^{-1} = \frac{1}{2} \lim_{\epsilon \to 0} \epsilon \int_{\mathbb{R}^n} |(x,\xi)|^{-1+\epsilon} \chi_L(\xi)\, d\xi,$$

where χ_L is the indicator function of L.

Comparing the latter equality with (1.9), we see that intersection bodies look very much like the unit balls of spaces embedding in L_{-1}, which provides motivation for introducing the concept of embedding in L_p with $p < 0$. To do this, we would like to extend the representation (1.9) to $p \le -1$. However, if $p \le -1$, the integral in the right-hand side of (1.9) diverges and we have to regularize it. The standard way of regularization is by using distributions. Applying both sides of (1.9) to a test function ϕ, we get

$$\langle \|x\|^p, \phi(x) \rangle = \int_{S^{n-1}} \langle |(x,\xi)|^p, \phi(x) \rangle \, d\mu(\xi).$$

Passing to Fourier transforms and using Lemmas 2.23 and 2.11, we get

$$\langle |(x,\xi)|^p, \phi(x) \rangle = \langle |t|^p, \int_{(x,\xi)=t} \phi(x)\, dx \rangle$$

$$= c_p \langle |z|^{-1-p}, \hat{\phi}(z\xi) \rangle.$$

If $-n < p < 0$, the distributions $\|\cdot\|^p$ on $\mathbb{R}^n$ and $|\cdot|^{-1-p}$ on $\mathbb{R}$ are locally integrable functions and, therefore, act by integration. We arrive at the following definition, first introduced in [K12]:

DEFINITION 1.6. *Let D be an origin-symmetric star body in $\mathbb{R}^n$, and let $X = (\mathbb{R}^n, \|\cdot\|_D)$. For $-n < p < 0$, we say that X embeds in L_p if there exists a finite Borel measure μ on S^{n-1} so that, for every even Schwartz test function $\phi \in \mathcal{S}(\mathbb{R}^n)$,*

$$\int_{\mathbb{R}^n} \|x\|_D^p \phi(x)\, dx = \int_{S^{n-1}} \left(\int_{\mathbb{R}} |z|^{-p-1} \hat{\phi}(z\theta)\, dt \right) d\mu(\theta).$$

A direct connection between intersection bodies and embeddings in negative L_p was established in [K14] (see Theorem 6.16):

THEOREM 1.7. *Let $1 \le k < n$. An origin-symmetric star body D in $\mathbb{R}^n$ is a k-intersection body if and only if the space $(\mathbb{R}^n, \|\cdot\|_D)$ embeds in L_{-k}. Both conditions are also equivalent to $\|\cdot\|_D^{-k}$ being a positive definite distribution.*

The advantage of this connection (and, consequently, of introducing embeddings in negative L_p) is that now one can try to extend to negative values of p different results about usual L_p-spaces. Every such extension gives new information about intersection bodies. Let us give several examples of this approach.

A well-known simple fact is that every two-dimensional normed space embeds in L_1 ([Her], [Fe], [Li]; see Corollary 6.8). How does this fact extend to embeddings in L_{-k}? It was proved in [K11, Th. 2] (see Corollary 4.9) that for every origin-symmetric convex body K in $\mathbb{R}^n$ and every $p \in [n-3, n)$ such that $p > 0$, the function $\|\cdot\|_K^{-p}$ represents a positive definite distribution. By Theorem 1.7 every n-dimensional Banach space embeds in L_{-n+3}. Putting $n = 2$, we get the property of two-dimensional spaces mentioned above. Putting $n = 4$, we see that every four-dimensional normed space embeds in L_{-1}. By Theorem 1.7, every four-dimensional origin-symmetric convex body is a 1-intersection body, which, by Lutwak's connection, solves in the affirmative the critical four-dimensional case of the Busemann-Petty problem.

Another well-known property of L_p-spaces, proved in [BDK], is that, for any $0 < p < q \le 2$, the space L_q embeds isometrically in L_p, so L_p-spaces become larger when p decreases from 2 (see Corollary 6.7). This result was extended to negative p in [K12, Th. 2] (see Theorem 6.17): every finite dimensional subspace of L_q, $0 < q \le 2$, embeds in L_{-p} for every $p \in (0, n)$. Hence, the unit ball of every n-dimensional subspace of L_q, $0 < q \le 2$, is a k-intersection body for every $k = 1, ..., n-1$. This gives plenty of examples of intersection bodies, and, in particular, proves a conjecture of M. Meyer [Me] that every polar projection body is an intersection body.

On the negative side, the solution to Schoenberg's problem (see Corollary 6.12) shows that for $q > 2$ and $n \ge 3$, the spaces ℓ_q^n do not embed in L_p with $0 < p \le 2$. This result was extended to negative p in [K8, Lemma 9] (Theorem 4.13): the spaces ℓ_q^n with $q > 2$ embed in L_{-p} only for $p \in [n-3, n)$. Putting $n = 5$, we see that ℓ_q^5 with $q > 2$ does not embed in L_{-1}, so the unit balls of these spaces are not 1-intersection bodies, which provides counterexamples to the Busemann-Petty problem in the critical dimension 5.

All these examples are isometric. An isomorphic example is given in Section 6.3. The factorization theorem of Maurey and Nikishin, [Ma], [Ni], implies that, for $0 < p < q < 1$, every Banach subspace of L_p is isomorphic to a subspace of L_q. It was proved in [KK2] that, for any $-\infty < p < q < 1$, $p \ne 0$, $q > 0$, every n-dimensional Banach subspace of L_p, $-n < p$, is isomorphic to a subspace of L_q with the Banach-Mazur distance depending only on p and q (see Theorem 6.26). In terms of intersection bodies this shows that origin-symmetric convex k-intersection bodies are isomorphically equivalent to unit balls of subspaces of L_q, $0 < q < 1$.

Chapter 6 is organized as follows. We start with several results on embedding of normed spaces in L_p with $p > 0$. In particular, we prove a Fourier analytic characterization of subspaces of L_p: if $p > 0$ is not an even integer, then a normed space $(\mathbb{R}^n, \|\cdot\|)$ embeds isometrically in L_p if and only if the Fourier transform of $\Gamma(-p/2)\|\cdot\|^p$ is a non-negative distribution outside of the origin in $\mathbb{R}^n$. We also prove a well-known old result (going back to P. Levy in the finite dimensional case and extended to the infinite dimensional case in [BDK]) that a finite dimensional normed space $(\mathbb{R}^n, \|\cdot\|)$ embeds isometrically in L_p with $0 < p \leq 2$ if and only if the function $\exp(-\|\cdot\|^p)$ is positive definite on $\mathbb{R}^n$.

In Section 6.2, we present a solution to Schoenberg's problem on positive definite functions. This problem was posed in 1938 in [Sc1] and was completely solved in the beginning of the 1990's in [Mis2] (for $q = \infty$) and [K1] (for $2 < q < \infty$): given $q > 2$, for which $p > 0$ is the function $\exp(-\|\cdot\|_q^p)$ positive definite, or, equivalently, for which $0 < p \leq 2$ does the space ℓ_q^n embed isometrically in L_p? As mentioned above, the answer is negative for $n \geq 3$, $q > 2$ and every $0 < p \leq 2$. If $n = 2$, the answer is affirmative if and only if $p \in [0, 1]$.

The concept of embedding in L_p with $p < 0$ and related results, described above, are presented in Section 6.3.

In Chapter 7, we consider the problem of finding the extremal sections of ℓ_q-balls, $0 < q \leq \infty$. In Section 7.1, we prove Ball's theorem on the maximal central hyperplane sections of the cube. The extremal sections of ℓ_q-balls with $0 < q \leq 2$ are considered in Section 7.2.

The Fourier analytic approach to sections was extended in [KRZ] to projections of convex bodies. The results for projections are surprisingly similar to those for sections. This similarity once again exposes the duality between sections and projections, which remains one of the most intriguing mysteries of convex geometry.

In place of the Fourier formula for sections (see formula (1.5)), we have a formula connecting the volume of hyperplane projections with the curvature function of the body, [KRZ, Th. 2]:

$$\widehat{f_L}(\theta) = -\pi \mathrm{Vol}_{n-1}\left(L\middle|\theta^{\perp}\right), \qquad \forall \theta \in S^{n-1}, \tag{1.10}$$

where $L\big|\theta^{\perp}$ is the projection of an origin-symmetric convex body L to the hyperplane $\theta^{\perp}$, and f_L is the curvature function of L extended from the sphere to the whole $\mathbb{R}^n$ as a homogeneous function of degree $-n-1$. We prove this formula in Section 8.1.

In Section 8.2, we prove a result of Barthe and Naor [BN] on the extremal hyperplane projections of ℓ_q-balls with $q > 2$: the minimal projection corresponds to the hyperplane orthogonal to the vector $(1, 0, ..., 0)$, and the maximal projection corresponds to the vector $(1, ..., 1)$.

In Section 8.3 we consider projection bodies that represent an analog of intersection bodies. An origin-symmetric convex body L in $\mathbb{R}^n$ is called a *projection body* if there exists another convex body K so that the support function of L in every direction is equal to the volume of the hyperplane projection of K to this direction: for every $\theta \in S^{n-1}$,

$$h_L(\theta) = \mathrm{Vol}_{n-1}(K\big|\theta^{\perp}).$$

The support function $h_L(\theta) = \max_{x \in L}(\theta, x)$ is equal to the dual norm $\|\theta\|_{L^*}$, where L^* stands for the polar body of L. We prove a Fourier analytic characterization of projection bodies. A convex body L is a projection body if and only if the Fourier transform of the support function h_L is a negative distribution outside of the origin. Consequently, by Theorem 1.4, if h_L is infinitely smooth on the sphere and n is an even integer, then L is a projection body if and only if, for every $\xi \in S^{n-1}$,

$$(-1)^{n/2} A^{(n)}_{L^*,\xi} \geq 0.$$

This result is very similar to the characterization of intersection bodies in Section 4.1 and explains the transition between the dimensions 2 and 3 in Shephard's problem.

In Section 8.4 we present a Fourier analytic solution to Shephard's problem, which is the projection counterpart of the Busemann-Petty problem. The problem reads as follows. Let K, L be convex origin-symmetric bodies in $\mathbb{R}^n$ and suppose that, for every $\theta \in S^{n-1}$,

$$\text{Vol}_{n-1}\left(K\middle|\theta^\perp\right) \leq \text{Vol}_{n-1}\left(L\middle|\theta^\perp\right). \tag{1.11}$$

Does it follow that

$$\text{Vol}_n(K) \leq \text{Vol}_n(L)? \tag{1.12}$$

The problem was solved independently by Petty [Pe] and Schneider [Sch1], who showed that the answer is affirmative if $n \leq 2$ and negative if $n \geq 3$. Our solution to the Shephard problem is based on formula (1.10) and a spherical version of Parseval's formula. This solution is in the spirit of the solution to the Busemann-Petty problem from Section 5.2.

The main goal of this work is to show connections between Fourier analysis and convex geometry. Our coverage of both subjects does not pretend to be complete; we emphasize the results that illustrate connections between the two areas. However, every chapter contains a section with historical remarks, open problems and other results on the subject that are not included in the main text. Many more results and references on sections and projections of convex bodies can be found in monographs and surveys [Ba5], [Ba6], [Ga3], [GiM], [KKo], [MiP], [MiS], [Sch3]. The book [Gr] contains numerous applications of spherical harmonics to convex geometry. The survey [K18] gives a quick introduction to Fourier methods for sections of convex bodies.

Acknowledgments. I would like to thank the referee for very useful suggestions about the contents and style of this book. Marina Yaskina and Artem Zvavitch helped me with figures. During the work on this book, I was partially supported by the US National Science Foundation, grant DMS-0136022.

CHAPTER 2

Basic Concepts

2.1. Star bodies

We say that a closed bounded set K in $\mathbb{R}^n$ is a *star body* if for every $x \in K$ each point of the interval $[0, x)$ is an interior point of K (in other words, every straight line passing through the origin crosses the boundary of K at exactly two points different from the origin), and the boundary of K is continuous in the sense that the *Minkowski functional* of K defined by

$$\|x\|_K = \min\{a \geq 0 : \ x \in aK\}$$

is a continuous function on $\mathbb{R}^n$.

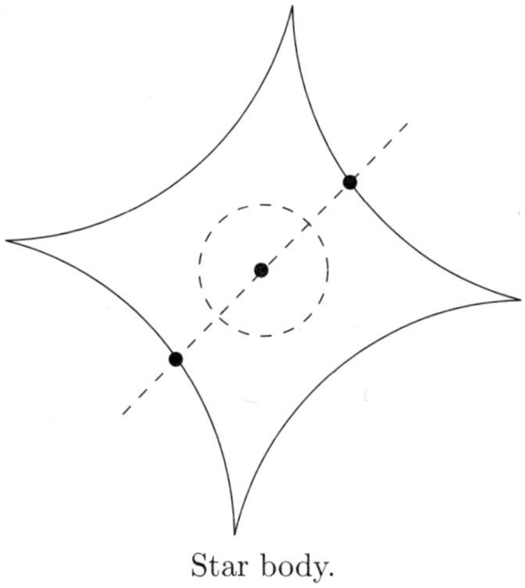

Star body.

It is easy to see that the Minkowski functional is a homogeneous function of degree 1 on $\mathbb{R}^n$ and that

$$K = \{x \in \mathbb{R}^n : \ \|x\|_K \leq 1\}.$$

Also, it follows from the definition that the origin is an interior point of every star body, so the Minkowski functional is strictly positive outside of the origin.

The *radial function* of a star body K is defined by

$$\rho_K(x) = \|x\|_K^{-1}, \qquad x \in \mathbb{R}^n.$$

If $x \in S^{n-1}$, then $\rho_K(x)$ is the "radius" of K in the direction of x, i.e. the distance from the origin to the boundary of K in the direction of x.

We define the *radial metric* on the set of all origin-symmetric star bodies in $\mathbb{R}^n$ by

$$\rho(K, L) = \max_{x \in S^{n-1}} |\rho_K(x) - \rho_L(x)|.$$

We use different summations of star bodies. If K and L are star bodies in $\mathbb{R}^n$, then *the Minkowski sum* of K and L is the set

$$K + L = \{x + y : \ x \in K, \ y \in L\},$$

where $x + y$ is the usual sum of vectors in $\mathbb{R}^n$.

If $p \in \mathbb{R} \setminus \{0\}$, the p-sum of K and L is a star body $K +_p L$ defined by

$$\|x\|_{K+_pL} = (\|x\|_K^p + \|x\|_L^p)^{1/p}$$

for every $x \in \mathbb{R}^n$. The case $p = -1$ corresponds to the *radial sum*, where the radius of the body $K +_{-1} L$ in every direction is equal to the sum of radii of K and L in this direction. The 0-sum (or multiplicative sum) $K +_0 L$ of two star bodies K and L is defined by

$$\|x\|_{K+_0L} = \sqrt{\|x\|_K \|x\|_L}.$$

A star body is called *k-smooth*, $k \in \mathbb{N} \cup \{0\}$, if the restriction of its Minkowski functional to the sphere S^{n-1} belongs to the space $C^k(S^{n-1})$ of k times continuously differentiable functions on the sphere S^{n-1}. Recall that the norm of the space C^k is defined as the maximum of sup-norms of the function itself and all its derivatives up to the order k. We say that a body is infinitely smooth if it is k-smooth for every $k \in \mathbb{N}$.

The condition that the origin is an interior point of a star body implies that certain negative powers of the Minkowski functional are locally integrable on $\mathbb{R}^n$. This simple fact is very important for us, since we are going to consider powers of the Minkowski functional as distributions, and local integrability implies that the action of these distributions on test functions is by integration; see Section 2.5.

LEMMA 2.1. *Let K be an origin-symmetric star body in $\mathbb{R}^n$. Then, for $0 < p < n$, the function $\|\cdot\|_K^{-p}$ is locally integrable on $\mathbb{R}^n$. Also, if f is a bounded integrable function on $\mathbb{R}^n$, then the function $\|\cdot\|_K^{-p} f(\cdot)$ is integrable on $\mathbb{R}^n$.*

PROOF. By the definition of a star body, K contains a Euclidean ball and is contained in another Euclidean ball with center at the origin. Therefore, there exist constants c_1, c_2 so that, for every $x \in \mathbb{R}^n$, $c_1|x|_2 \leq \|x\|_K \leq c_2|x|_2$, where $|\cdot|_2$ is the Euclidean norm in $\mathbb{R}^n$.

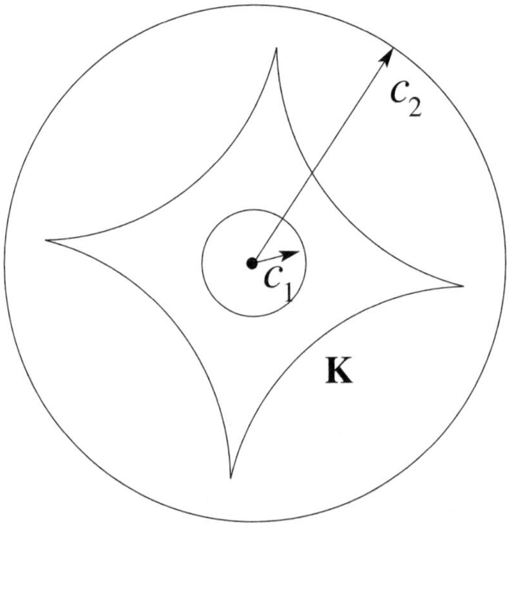

On the other hand, the function $|\cdot|_2^{-p}$ is integrable on any Euclidean ball with center at the origin, which follows from an elementary calculation in polar coordinates. This proves the first statement, and the second easily follows. In fact, on any Euclidean ball with center at the origin the function $\|\cdot\|_K^{-p}$ is locally integrable and f is bounded, while outside of this ball f is integrable and $\|\cdot\|_K^{-p}$ is bounded. □

For $\xi \in S^{n-1}$, we define the *parallel section function* of K in the direction of ξ as a function on $\mathbb{R}$ given by

$$A_{K,\xi}(t) = \mathrm{Vol}_{n-1}(K \cap \{\xi^\perp + t\xi\}),$$

where $\{\xi^\perp + t\xi\}$ is the hyperplane perpendicular to ξ at distance t from the origin. We use the notation A_ξ when it is clear what body is considered.

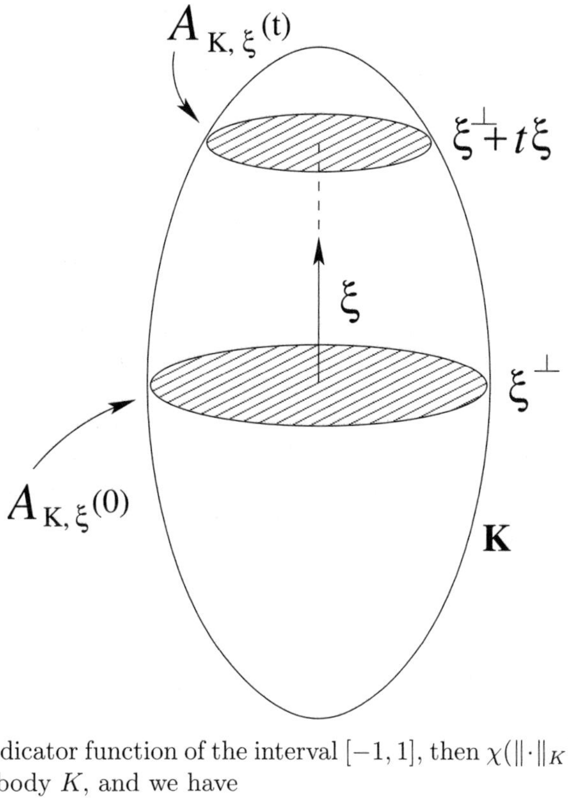

If χ is the indicator function of the interval $[-1,1]$, then $\chi(\|\cdot\|_K)$ is the indicator function of the body K, and we have

$$A_{K,\xi}(t) = \int_{(x,\xi)=t} \chi(\|x\|_K)\, dx. \tag{2.1}$$

For $t = 0$, writing the integral in the right-hand side in the polar coordinates of the hyperplane $(x,\xi) = 0$, we get the *polar formula for the volume of sections* (note that the dimension of a hyperplane is $n-1$, which explains the power r^{n-2}):

$$A_{K,\xi}(0) = \mathrm{Vol}_{n-1}(K \cap \xi^\perp) = \int_{(x,\xi)=0} \chi(\|x\|_K)\, dx$$

$$= \int_{S^{n-1}\cap\xi^\perp} \left(\int_0^\infty r^{n-2}\chi(r\|\theta\|_K)\, dr \right) d\theta$$

$$= \int_{S^{n-1}\cap\xi^\perp} \left(\int_0^{1/\|\theta\|_K} r^{n-2}\, dr \right) d\theta$$

$$= \frac{1}{n-1} \int_{S^{n-1}\cap\xi^\perp} \|\theta\|_K^{-n+1}\, d\theta$$

$$= \frac{1}{n-1} \int_{S^{n-1}\cap\xi^\perp} \rho_K^{n-1}(\theta)\, d\theta. \tag{2.2}$$

If H is an m-dimensional subspace of $\mathbb{R}^n$, then the m-dimensional volume of the section of K by H can be expressed in terms of the Minkowski functional in a similar way:

$$\mathrm{Vol}_m(K \cap H) = \frac{1}{m} \int_{S^{n-1}\cap H} \|\theta\|_K^{-m}\, d\theta. \tag{2.3}$$

In particular, putting $m = n$,

$$\mathrm{Vol}_n(K) = \frac{1}{n} \int_{S^{n-1}} \|\theta\|_K^{-n} d\theta = \frac{1}{n} \int_{S^{n-1}} \rho_K^n(\theta) d\theta. \tag{2.4}$$

2.2. Convex bodies

Our main references for results in convex geometry are the books by Schneider [Sch3] and Gardner [Ga3].

A closed compact set K in $\mathbb{R}^n$ with non-empty interior is called a *convex body* if it contains the line segment connecting any two of its points, i.e. for every $x, y \in K$ and $\lambda \in [0,1]$, the point $\lambda x + (1-\lambda)y \in K$.

If a convex body K is origin-symmetric, then its Minkowski functional is a norm on $\mathbb{R}^n$. In fact, as it is for any origin-symmetric star body, this functional is even 1-homogeneous, i.e. for every $x \in \mathbb{R}^n$ and $t \in \mathbb{R}$, $\|tx\|_K = |t|\|x\|_K$. Also, since K is origin-symmetric and its interior in non-empty, the origin is an interior point of K, which implies that $\|x\|_K = 0$ if and only if $x = 0$. The third defining property of a norm, the triangle inequality

$$\|x + y\|_K \le \|x\|_K + \|y\|_K,$$

follows from convexity of K and is easy to verify.

One of the main tools of convex geometry is the Brunn-Minkowski inequality. This inequality was first proved for three-dimensional convex bodies by Brunn in 1887 and for convex bodies in arbitrary dimension by Minkowski in 1910. Lusternik [Lus] proved the general Brunn-Minkowski inequality for arbitrary non-empty bounded measurable sets. There exist many different proofs of this inequality; see [Sch3, p. 309], [Ga4], [Ba5]. The short proof that we present here is due to Hadwiger and Ohmann [HO].

THEOREM 2.2. *(Brunn-Minkowski Inequality) If K and L are non-empty compact sets in $\mathbb{R}^n$, then for every $\lambda \in [0,1]$,*

$$\mathrm{Vol}_n(\lambda K + (1-\lambda)L)^{1/n} \ge \lambda \mathrm{Vol}_n(K)^{1/n} + (1-\lambda)\mathrm{Vol}_n(L)^{1/n}. \tag{2.5}$$

PROOF. Since $\mathrm{Vol}_n(\lambda K) = \lambda^n \mathrm{Vol}_n(K)$, it is enough to prove that

$$\mathrm{Vol}_n(K + L)^{1/n} \ge \mathrm{Vol}_n(K)^{1/n} + \mathrm{Vol}_n(L)^{1/n}. \tag{2.6}$$

First, consider the case where

$$K = \prod_{i=1}^{n}[a_i, b_i] \quad \text{and} \quad L = \prod_{i=1}^{n}[c_i, d_i]$$

are rectangular parallelepipeds in $\mathbb{R}^n$ with sides parallel to the coordinate hyperplanes. Denote by $x_i = b_i - a_i$, $y_i = d_i - c_i$ the lengths of the sides of K and L, respectively. Then

$$K + L = \prod_{i=1}^{n}[a_i + c_i, b_i + d_i]$$

and

$$\mathrm{Vol}_n(K+L) = \prod_{i=1}^{n}(x_i + y_i), \ \mathrm{Vol}_n(K) = \prod_{i=1}^{n} x_i, \ \mathrm{Vol}_n(L) = \prod_{i=1}^{n} y_i.$$

By the arithmetic-geometric mean inequality,

$$\left(\prod_{i=1}^{n}\frac{x_i}{x_i+y_i}\right)^{1/n} + \left(\prod_{i=1}^{n}\frac{y_i}{x_i+y_i}\right)^{1/n} \leq \frac{1}{n}\sum_{i=1}^{n}\frac{x_i}{x_i+y_i} + \frac{1}{n}\sum_{i=1}^{n}\frac{y_i}{x_i+y_i} = 1.$$

This proves inequality (2.6) for two rectangular boxes.

Now suppose that K is the union of m_1 disjoint (up to a set of volume zero) rectangular boxes and L is the union of m_2 disjoint rectangular boxes. We prove inequality (2.6) for such K and L by induction on the number $m_1 + m_2$. Since the result is proved in the case $m_1 + m_2 = 2$, we can assume that $m_1 + m_2 > 2$ and that K contains at least two disjoint boxes.

Note that both sides of inequality (2.6) do not change if we translate any of the sets K or L. Translating K if necessary, we can place it in such a way that at least two boxes in K are separated from each other by a coordinate plane, say $\{x_n = 0\}$. Denote by K_+ and K_- the unions of boxes in $K \cap \{x_n \geq 0\}$ and $K \cap \{x_n \leq 0\}$, respectively. Clearly, the number of boxes in each of K_+ and K_- is smaller than m_1.

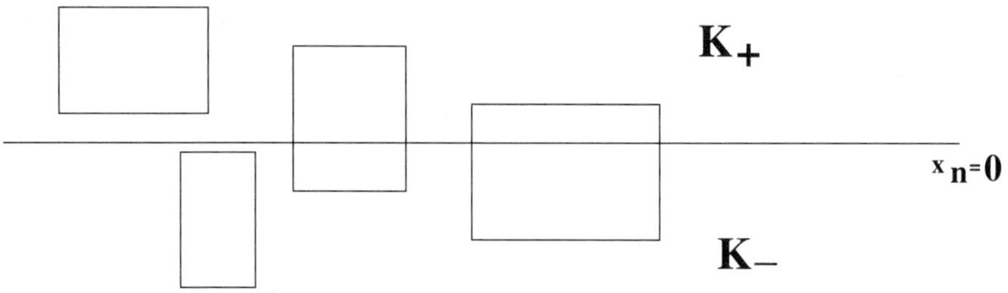

Now translate L so that

$$\frac{\mathrm{Vol}_n(K_+)}{\mathrm{Vol}_n(K)} = \frac{\mathrm{Vol}_n(L_+)}{\mathrm{Vol}_n(L)} \quad \text{and} \quad \frac{\mathrm{Vol}_n(K_-)}{\mathrm{Vol}_n(K)} = \frac{\mathrm{Vol}_n(L_-)}{\mathrm{Vol}_n(L)},$$

where L_+ and L_- are the parts of L contained in $\{x_n \geq 0\}$ and $\{x_n \leq 0\}$ (the translation must provide one of these equalities; the second will hold automatically, since the sum of the left- and right-hand sides is 1).

The sets $K_+ + L_+$ and $K_- + L_-$ are disjoint (up to a set of volume zero) subsets of $K + L$, since they are on different sides of the hyperplane $\{x_n = 0\}$. Besides, the sum of numbers of boxes in K_+ and L_+ is smaller than $m_1 + m_2$, and the same is true for K_- and L_-. By the induction hypothesis, inequality (2.6) holds for each of these two pairs of sets. We get

$$\mathrm{Vol}_n(K+L) \geq \mathrm{Vol}_n(K_+ + L_+) + \mathrm{Vol}_n(K_- + L_-)$$

$$\geq \left(\mathrm{Vol}_n(K_+)^{1/n} + \mathrm{Vol}_n(L_+)^{1/n}\right)^n + \left(\mathrm{Vol}_n(K_-)^{1/n} + \mathrm{Vol}_n(L_-)^{1/n}\right)^n$$

$$= \mathrm{Vol}_n(K_+)\left(1 + \frac{\mathrm{Vol}_n(L)^{1/n}}{\mathrm{Vol}_n(K)^{1/n}}\right)^n + \mathrm{Vol}_n(K_-)\left(1 + \frac{\mathrm{Vol}_n(L)^{1/n}}{\mathrm{Vol}_n(K)^{1/n}}\right)^n$$

$$= \left(\mathrm{Vol}_n(K)^{1/n} + \mathrm{Vol}_n(L)^{1/n}\right)^n.$$

We have proved (2.6) for finite unions of disjoint boxes, and now we can get the general result approximating compact sets by such unions. □

It follows from the Brunn-Minkowski inequality that if K is an origin-symmetric convex body in $\mathbb{R}^n$, then the central hyperplane section has maximal $(n-1)$-dimensional volume among all hyperplane sections orthogonal to a given direction.

THEOREM 2.3. *(Brunn's Theorem) Let K be a convex body in $\mathbb{R}^n$. Fix $\xi \in S^{n-1}$ and let $A_{K,\xi}$ be the parallel section function in the direction of ξ. Then the following hold.*
(i) The function $A_{K,\xi}^{1/(n-1)}$ is concave on its support.
(ii) If K is origin-symmetric, then, for each $\xi \in S^{n-1}$ and $t \in \mathbb{R}$,

$$A_{K,\xi}(t) \leq A_{K,\xi}(0) = \mathrm{Vol}_{n-1}(K \cap \xi^\perp),$$

i.e. the central section has maximal $(n-1)$-dimensional volume among all hyperplane sections of K perpendicular to a given direction.
(iii) If K is origin-symmetric and 2-smooth, then the second derivative $A''_{K,\xi}(0) \leq 0$.

PROOF. For every $t \in \mathbb{R}$, consider the compact convex set

$$K_t = K \cap \{x \in \mathbb{R}^n : \ (x,\xi) = t\}.$$

By the convexity of K, for any $\lambda \in [0,1]$ and any $s, t \in \mathbb{R}$ such that the sets K_s and K_t are non-empty (i.e. s and t belong to the support of the function $A_{K,\xi}$), we have

$$\lambda K_s + (1-\lambda) K_t \subset K_{\lambda s + (1-\lambda)t},$$

and therefore

$$\lambda P(K_s) + (1-\lambda) P(K_t) \subset P(K_{\lambda s + (1-\lambda)t}),$$

where P is the orthogonal projection to the hyperplane $(x,\xi) = 0$, and the addition of sets is considered in the hyperplane $(x,\xi) = 0$.

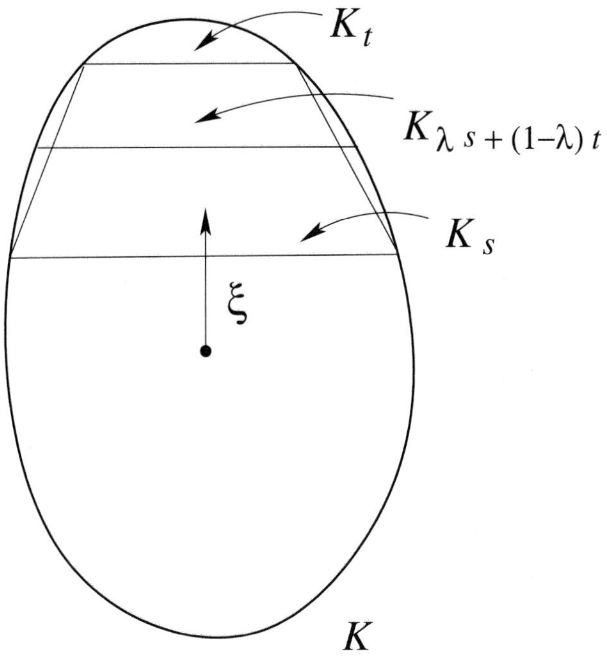

The orthogonal projections have the same $(n-1)$-dimensional volume as the original sets. By the Brunn-Minkowski inequality applied to the compact convex sets $P(K_s)$ and $P(K_t)$ in $\mathbb{R}^{n-1} = \xi^\perp$,

$$\lambda(\mathrm{Vol}_{n-1}(K_s))^{1/(n-1)} + (1-\lambda)(\mathrm{Vol}_{n-1}(K_t))^{1/(n-1)}$$
$$= \lambda(\mathrm{Vol}_{n-1}(P(K_s)))^{1/(n-1)} + (1-\lambda)(\mathrm{Vol}_{n-1}(P(K_t)))^{1/(n-1)}$$
$$\leq (\mathrm{Vol}_{n-1}(\lambda P(K_s) + (1-\lambda)P(K_t)))^{1/(n-1)}$$
$$\leq (\mathrm{Vol}_{n-1}(P(K_{\lambda s+(1-\lambda)t})))^{1/(n-1)} = (\mathrm{Vol}_{n-1}(K_{\lambda s+(1-\lambda)t}))^{1/(n-1)}.$$

Since $\mathrm{Vol}_{n-1}(K_s) = A_{K,\xi}(s)$, we have proved that

$$\lambda A_{K,\xi}^{1/(n-1)}(s) + (1-\lambda)A_{K,\xi}^{1/(n-1)}(t) \leq A_{K,\xi}^{1/(n-1)}(\lambda s + (1-\lambda)t),$$

so the function $A_{K,\xi}^{1/(n-1)}$ is concave on its support. A concave even function has maximum at zero, which proves the second statement. If, in addition, K is 2-smooth, then, as we show below in Lemma 2.4, the function $A_{K,\xi}$ is twice differentiable in a neighborhood of zero, which implies the third statement of the theorem. □

We have just used the fact that the parallel section function of a smooth body is also smooth. We need this fact in a general form. We say that a family of functions is uniformly differentiable if convergence in the limits defining the derivatives is uniform with respect to this family of functions.

LEMMA 2.4. *Let K be an m-smooth origin-symmetric convex body in $\mathbb{R}^n$, where $m \in \mathbb{N} \cup \{0\}$. Then for all $\xi \in S^{n-1}$ the parallel section functions $A_{K,\xi}$ are (uniformly with respect to ξ) m times continuously differentiable in some neighborhood U of zero. Moreover, for every fixed $t \in U$, the derivative $A_{K,\xi}^{(m)}(t)$ is a continuous function of the variable ξ on the sphere.*

PROOF. Fix a direction $\xi \in S^{n-1}$. For every t consider the hyperplane $H_t = \{x : (x,\xi) = t\}$. Since the interior of K is non-empty and K is origin-symmetric, it contains a Euclidean ball with radius r and center at the origin. We assume that $|t| < r$. Let S^{n-2} be the unit Euclidean sphere in H_t with center at the orthogonal projection of the origin to H_t (which is inside of $K \cap H_t$ since $|t| < r$), and let $\rho_{K\cap H_t}$ be the radial function of the body $K \cap H_t$ in $(H_t, S^{n-2}) = \mathbb{R}^{n-1}$ with respect to this center. By the polar formula for the volume of sections (see (2.4)),

$$A_{K,\xi}(t) = \mathrm{Vol}_{n-1}(K \cap H_t) = \frac{1}{n-1}\int_{S^{n-2}} \rho^{n-1}_{K\cap H_t}(\theta)\, d\theta. \tag{2.7}$$

The question is reduced to uniform differentiability (with respect to θ and ξ) of $\rho_{K\cap H_t}(\theta)$ by the variable t.

Fix θ and consider the two-dimensional plane passing through the origin and spanned by ξ and θ. Let D be the section of K by this plane, and let ρ_D be the radial function of D defined on $[0, 2\pi]$. Since K is C^m, the function ρ_D is C^m. Let us now get an implicit formula expressing $\rho_{K\cap H_t}(\theta)$ in terms of ρ_D. The line connecting the origin with the point on the boundary corresponding to $\rho_{K\cap H_t}(\theta)$ is at the angle $\phi = \arctan(\frac{t}{\rho_{K\cap H_t}(\theta)})$ with respect to the hyperplane $\xi^\perp$. Therefore, the radius at this point is equal to $\rho_D(\arctan(\frac{t}{\rho_{K\cap H_t}(\theta)}))$. From the right triangle,

$$\rho_{K\cap H_t}(\theta) = \sqrt{\rho_D^2(\arctan(\frac{t}{\rho_{K\cap H_t}(\theta)})) - t^2}. \tag{2.8}$$

This is an implicit formula that we are now going to use to compute the derivatives of $y(t) = \rho_{K\cap H_t}(\theta)$. Formula (2.8) now has the form

$$y = \sqrt{\rho_D^2(\arctan(\frac{t}{y})) - t^2}.$$

Differentiating both sides by t and separating the derivative of y, we get

$$y'(t) = \frac{\rho_D(\arctan(t/y))\rho_D'(\arctan(t/y))\frac{1}{y^2+t^2}(y - ty'(t)) - t}{y};$$

therefore

$$y'(t) = \frac{\rho_D(\arctan(\frac{t}{y}))\rho_D'(\arctan(\frac{t}{y}))(1/(y^2+t^2))y - t}{y + t\rho_D(\arctan(\frac{t}{y}))\rho_D'(\arctan(\frac{t}{y}))\frac{1}{y^2+t^2}}.$$

Again, K contains a Euclidean ball with center at the origin, so ρ_D and y are separated from zero for small t with a bound independent on θ and ξ. Also, $\rho_K \in C^m(S^{n-1})$, so ρ_D and all its derivatives up to the order m are bounded from above uniformly with respect to θ and ξ. This shows that the denominator in the expression for y' is positive for small t, so the derivative exists in a neighborhood of zero uniformly with respect to θ and ξ. Computing derivatives of higher orders is similar (note that we never get zero in the denominator if t is small). This shows that y is m times uniformly differentiable in some neighborhood of zero.

It follows from (2.7) that uniform differentiability of $y = \rho_{K\cap H_t}(\theta)$ implies that $A_{K,\xi}(t)$ is uniformly (with respect to ξ) differentiable in a neighborhood of zero. This also implies the last statement of the lemma by induction by m. Given the result for $m-1$, we use the fact that, for every t in some neighborhood of zero, the function $\xi \mapsto A^{(m)}_{K,\xi}(t)$ is the limit in $C(S^{n-1})$ (i.e. uniformly by ξ) of continuous functions

$$\xi \mapsto \frac{A^{(m-1)}_{K,\xi}(t+h) - A^{(m-1)}_{K,\xi}(t)}{h},$$

as $h \to 0$. □

Another application of the Brunn-Minkowski inequality is the following result of Busemann [Bu1] (a proof can be found in [Ga3, Th. 8.1.10] or [MiP, Th. 3.9]; the result in [MiP] is more general):

THEOREM 2.5. *(Busemann's Theorem) Suppose that K is an origin-symmetric convex body in $\mathbb{R}^n$. For each $\xi \in S^{n-1}$, let $r(\xi)$ be the $(n-1)$-dimensional volume of the central hyperplane section of K perpendicular to ξ. Then the body whose radius in each direction ξ is $r(\xi)$ is itself convex.*

We say that a compact set K with volume 1 in $\mathbb{R}^n$ is in *isotropic position* if for each $\xi \in S^{n-1}$

$$\int_K (x,\xi)^2 dx = L_K^2, \tag{2.9}$$

where L_K is a constant that is called the *isotropic constant* of K. Since, for any compact set K, the left-hand side of (2.9) is a positive quadratic form of ξ, there exists a linear transformation T of $\mathbb{R}^n$ so that TK is in isotropic position; see [Ba7] for details. Note that the unit ball of any normed space with a symmetric basis is in isotropic position after a dilation making the volume equal to 1 (if we expand $(x,\xi)^2$, the terms with $x_i x_j$, $i \neq j$, integrate to zero because of symmetry, and the terms with x_i^2 generate equal integrals). In particular, if we dilate the unit balls of the spaces ℓ_p^n, $0 < p \le \infty$, so that the volumes of the dilated balls are equal to 1, then we get bodies that are in isotropic position.

Hensley [He2] has proved (see Corollary 2.7 below) that there exist absolute constants a_1, a_2 so that for any symmetric convex body K in $\mathbb{R}^n$ in isotropic position and any $\xi_1, \xi_2 \in S^{n-1}$,

$$a_1 \le \frac{\mathrm{Vol}_{n-1}(K \cap \xi_1^\perp)}{\mathrm{Vol}_{n-1}(K \cap \xi_2^\perp)} \le a_2.$$

The proof that we present here is due to Bourgain [Bo3]. For other proofs, see [MiP] or [Ba7].

LEMMA 2.6. *Let K be an origin-symmetric convex body in $\mathbb{R}^n$, $n \ge 2$, with $\mathrm{Vol}_n(K) = 1$. There exist absolute (not depending on K, n) constants $0 < c, d < \infty$ so that for any $\xi \in S^{n-1}$ and $t \in \mathbb{R}$,*

$$A_{K,\xi}(t) \le c A_{K,\xi}(0) \exp(-d A_{K,\xi}(0)|t|),$$

where $A_{K,\xi}$ is the parallel section function of K in the direction of ξ.

PROOF. Fix ξ and let $\phi(t) = A_{K,\xi}(t)$. We can assume that

$$|t|\phi(0) > 20. \tag{2.10}$$

In fact, by Brunn's theorem, Theorem 2.3, $\phi(0) > \phi(t)$, so if $|t|\phi(0) < 20$, then we can put $d = 1, c = e^{20}$ to get the required estimate.

If $\phi(t) \neq 0$, then $K \cap \{(x,\xi) = t\} \neq \emptyset$, and, since K is convex, there exists a cone with the base $K \cap \xi^\perp$ and height $|t|$ which is inside of K, so the volume of this cone is less than or equal to the volume of K :

$$1 = \mathrm{Vol}_n(K) \geq \frac{1}{n}|t|\phi(0). \tag{2.11}$$

Therefore, we can assume that

$$\phi(t) \geq \phi(0)e^{-n} \tag{2.12}$$

because, otherwise, by (2.11),

$$\phi(t) \leq \phi(0)e^{-n} \leq \phi(0)\exp(-|t|\phi(0)),$$

and we get the required estimate again.

For every $t \in \mathbb{R}$ and $\epsilon > 0$, we have (by Theorem 2.3 the function ϕ is decreasing from zero)

$$1 = \mathrm{Vol}_n(K) > \int_0^\infty \phi(s)\ ds \geq \int_0^{\epsilon|t|} \phi(s)\ ds \geq \epsilon|t|\phi(\epsilon|t|). \tag{2.13}$$

Again by Brunn's theorem, Theorem 2.3, the function $\phi(\cdot)^{1/(n-1)}$ is concave, so

$$\begin{aligned}\phi(\epsilon|t|) &\geq \left((1-\epsilon)\phi(0)^{1/(n-1)} + \epsilon\phi(|t|)^{1/(n-1)}\right)^{n-1}\\ &= \phi(0)\left(1-\epsilon+\epsilon\left(\frac{\phi(t)}{\phi(0)}\right)^{1/(n-1)}\right)^{n-1}\\ &= \phi(0)\left(1-\epsilon+\epsilon\exp(\ln(\alpha)/(n-1))\right)^{n-1},\end{aligned}$$

where $\alpha = \phi(t)/\phi(0)$. Combining the latter inequality with (2.13) and using an elementary inequality $\exp(x) \geq 1 + x$, we get

$$\begin{aligned}1 \geq \epsilon|t|\phi(\epsilon|t|) &\geq \epsilon|t|\phi(0)\left(1-\epsilon+\epsilon(1+\frac{\ln\alpha}{n-1})\right)^{n-1}\\ &= \epsilon|t|\phi(0)\left(1+\frac{\epsilon\ln\alpha}{n-1}\right)^{n-1}.\end{aligned}$$

Suppose now that $\epsilon < 1/4$. Note that $\alpha \leq 1$, so $\ln\alpha \leq 0$. By (2.12), $\ln\alpha > -n$, so $\epsilon\ln\alpha/(n-1) > -1/2$. The function $\ln(1+x)/x$ is bounded on the interval $[-1/2, 0)$, so there exists an absolute constant A so that

$$\ln\left(1+\frac{\epsilon\ln\alpha}{n-1}\right) \geq A\frac{\epsilon\ln\alpha}{n-1}.$$

Therefore, we have

$$1 \geq \epsilon|t|\phi(0)\left(\frac{\phi(t)}{\phi(0)}\right)^{A\epsilon}.$$

In the latter inequality, we can put

$$\epsilon = \frac{e}{|t|\phi(0)},$$

because, by (2.10), this ϵ is less than $1/4$. We get

$$\phi(t) \leq \phi(0)\exp(-\frac{|t|\phi(0)}{Ae}).$$

We have proved the desired inequality with $c = 1$ and $d = 1/Ae$. Now adjust c, d to accommodate the assumptions (2.10) and (2.12). □

We now prove the result of Hensley.

COROLLARY 2.7. *There exist absolute constants a_1, a_2 so that for any origin-symmetric convex body K in isotropic position we have*

$$\frac{a_1}{L_K} \leq \mathrm{Vol}_{n-1}(K \cap \xi^\perp) \leq \frac{a_2}{L_K}, \quad \forall \xi \in S^{n-1}.$$

PROOF. Note that the condition $\mathrm{Vol}_n(K) = 1$ is included in the definition of isotropic position. By the Fubini theorem,

$$L_K^2 = \int_K (x,\xi)^2 dx = \int_{\mathbb{R}} t^2 A_{K,\xi}(t)\, dt.$$

Using Brunn's theorem, Theorem 2.3, for any $\xi \in S^{n-1}$,

$$1 = \mathrm{Vol}_n(K) = \int_{\mathbb{R}} A_{K,\xi}(t)\, dt$$

$$\leq \int_{-2L_K}^{2L_K} A_{K,\xi}(0)\, dt + \int_{|t|>2L_K} \frac{t^2}{4L_K^2} A_{K,\xi}(t)\, dt$$

$$\leq 4L_K A_{K,\xi}(0) + \frac{1}{4L_K^2} \int_{\mathbb{R}} t^2 A_{K,\xi}(t)\, dt = 4L_K A_{K,\xi}(0) + \frac{1}{4},$$

so $L_K A_{K,\xi}(0) > 3/16$.

On the other hand, by Lemma 2.6,

$$L_K^2 = \int_{\mathbb{R}} t^2 A_{K,\xi}(t)\, dt$$

$$\leq c A_{K,\xi}(0) \int_{\mathbb{R}} t^2 \exp(-d|t| A_{K,\xi}(0))\, dt = \frac{C}{A_{K,\xi}(0)^2},$$

where C is an absolute constant. □

Definitions and results in the rest of this section will be used in the study of projections of convex bodies in Chapter 8.

If K and L are two convex bodies in $\mathbb{R}^n$, the *mixed volume* $V_1(K, L)$ is equal to

$$V_1(K,L) = \frac{1}{n} \lim_{\epsilon \to +0} \frac{\mathrm{Vol}_n(K + \epsilon L) - \mathrm{Vol}_n(K)}{\epsilon};$$

see [Ga3, p. 354] or [Sch3, p. 272]. Using the Brunn-Minkowski inequality, one can easily prove the so-called first Minkowski inequality (see [Ga3, p. 369]):

THEOREM 2.8. *(First Minkowski Inequality) For any convex bodies K, L in $\mathbb{R}^n$,*

$$V_1(K,L) \geq \mathrm{Vol}_n(K)^{(n-1)/n} \mathrm{Vol}_n(L)^{1/n}.$$

PROOF. By the Brunn-Minkowski inequality (2.5),

$$nV_1(K,L) \geq \lim_{\epsilon \to +0} \frac{(\mathrm{Vol}_n(K)^{1/n} + \epsilon \mathrm{Vol}_n(L)^{1/n})^n - \mathrm{Vol}_n(K)}{\epsilon}.$$

Computing the limit, we get the result. □

The *support function* of a convex body K in $\mathbb{R}^n$ is defined by

$$h_K(x) = \max_{\{\xi \in \mathbb{R}^n : \|\xi\|_K = 1\}} (x, \xi), \quad x \in \mathbb{R}^n.$$

If K is origin-symmetric, then h_K is a norm on $\mathbb{R}^n$. The convex body K^* which is the unit ball of this norm is called the *polar body* of K:

$$h_K(x) = \|x\|_{K^*}, \qquad x \in \mathbb{R}^n.$$

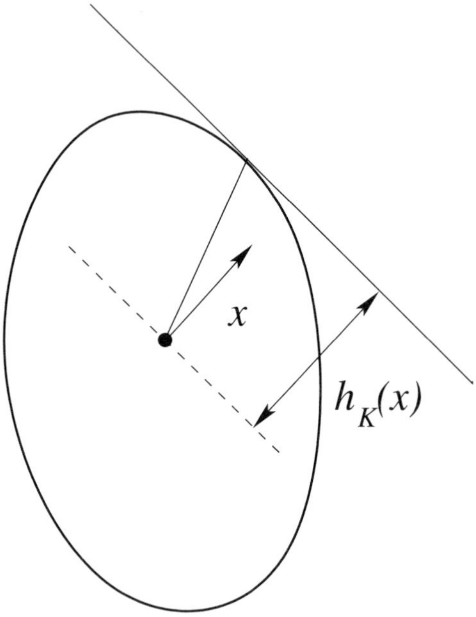

The support function h_K.

The *surface area measure* $S(K, \cdot)$ of a convex body K in $\mathbb{R}^n$ is defined as follows: for every Borel set $E \subset S^{n-1}$, $S(K, E)$ is equal to Lebesgue measure of the part of the boundary of K where normal vectors belong to E (see, for example, [Ga3, p. 351]). The Minkowski existence theorem shows that one gets essentially every symmetric measure on S^{n-1} as the surface area measure of some convex body:

THEOREM 2.9. *(Minkowski Existence Theorem) A finite Borel measure μ on the sphere S^{n-1} is the surface area measure of some convex body K in $\mathbb{R}^n$ if and only if μ is not supported in any great subsphere and the vector-valued integral*

$$\int_{S^{n-1}} u \, d\mu(u) = 0.$$

For a proof of Theorem 2.9, see [Sch3, p. 389]. At first glance, the result of this theorem looks surprising, because the condition on the measure is too general. To illustrate this theorem, consider the case where $n = 2$ and μ is the sum of m atoms at the points $u_1, ..., u_m$ with weights $w_1, ..., w_m$. The condition of the theorem is that

$$\sum_{i=1}^{m} w_i u_i = 0.$$

Let v_i be vectors obtained from u_i by rotation by the angle $\pi/2$. Clearly,

$$\sum_{i=1}^{m} w_i v_i = 0.$$

Adding vectors $w_i v_i$ geometrically (starting every next vector from the endpoint of the previous one), we can get a polygon whose surface area measure is exactly μ.

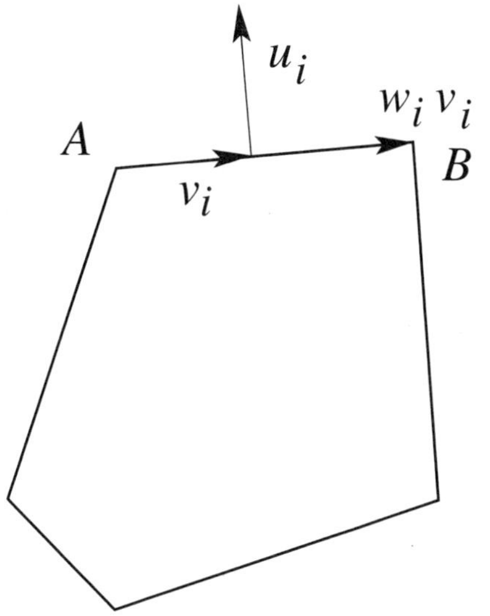

The well-known *Cauchy projection formula* ([Ga3, p. 361]) expresses the volume of projections of the body K in terms of the surface area measure:

$$\text{Vol}_{n-1}\left(K\middle|\theta^{\perp}\right) = \frac{1}{2}\int_{S^{n-1}} |(\theta, v)| dS(K, v), \qquad \theta \in S^{n-1}. \tag{2.14}$$

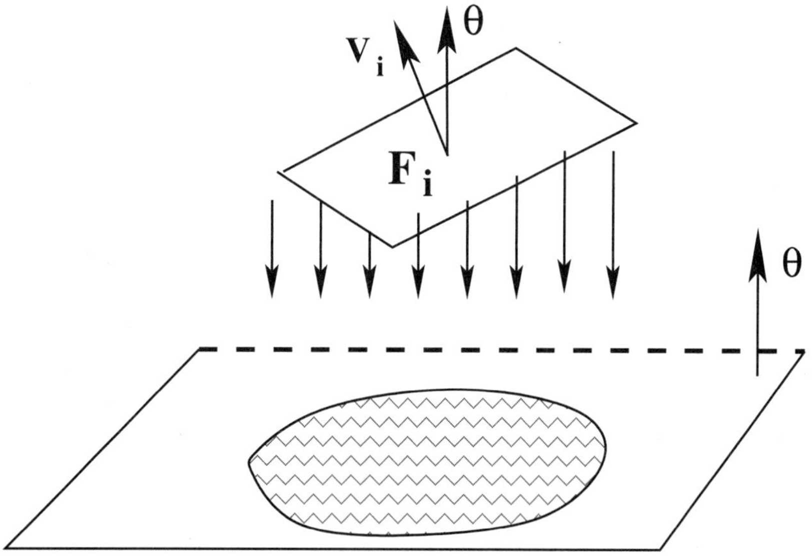

To prove this formula, first consider the case where K is a polytope with faces $F_1, ..., F_k$. The surface area measure is then an atomic measure on S^{n-1}, supported in the points $v_1, ..., v_k$ corresponding to unit normal vectors to all faces, with weights $\mathrm{Vol}_{n-1}(F_i) = S(K, \{v_i\})$. Then the projection of each face to the hyperplane $\theta^\perp$ is equal to the cosine of the angle between the normal vector v_i and the direction θ, which is $|(v_i, \theta)|$, times the area of the face, which is equal to $S(K, \{v_i\})$. Taking the sum, we get the integral in the Cauchy formula. The coefficient $1/2$ is necessary because we cover each point of the projection twice. The general case follows by approximation.

The mixed volume can be expressed in terms of the support functions and surface area measures, [Sch3, p. 275]:

$$V_1(K, L) = \frac{1}{n} \int_{S^{n-1}} h_L(x) dS(K, x). \tag{2.15}$$

To see this, again consider the case where K is a polytope. When we add ϵL to K, each face F_i of K moves outward in the direction of its normal vector v_i by $\epsilon h_L(v_i)$. Then the volume changes by the area of F_i (which is $S(K, v_i)$) times $\epsilon h_L(v_i)$. Adding these changes, we get the integral in (2.15) times ϵ. Other changes of the volume are at the rate of ϵ^2 or smaller.

We often consider bodies with absolutely continuous surface area measures. A convex body K is said to have the *curvature function*

$$f_K : S^{n-1} \to \mathbb{R}$$

if its surface area measure $S(K, \cdot)$ is absolutely continuous with respect to Lebesgue measure σ_{n-1} on S^{n-1} and if

$$\frac{dS(K, \cdot)}{d\sigma_{n-1}} = f_K \in L_1(S^{n-1}).$$

The curvature function f_K is the reciprocal Gauss curvature, viewed as a function of the unit normal vector (see [Sch3, p. 419]).

In the case where K has the curvature function formula (2.15) becomes

$$V_1(K, L) = \frac{1}{n} \int_{S^{n-1}} h_L(x) f_K(x) dx.$$

If $K = L$, we have $V_1(K, L) = \mathrm{Vol}_n(K)$, so

$$\mathrm{Vol}_n(K) = \frac{1}{n} \int_{S^{n-1}} h_K(x) f_K(x) dx. \tag{2.16}$$

It is well known that one can approximate any convex body in $\mathbb{R}^n$ in the radial metric by a sequence of infinitely smooth convex bodies. This can be proved by a simple convolution argument (see [Sch3, Th. 3.3.1] for the proof):

THEOREM 2.10. *Let K be an origin-symmetric convex body with support function h_K and containing a Euclidean ball with center at the origin and radius $1/C$. For $\epsilon > 0$, let $\phi_\epsilon : [0, \infty) \to [0, \infty)$ be an infinitely differentiable function supported in $[\epsilon/2, \epsilon]$ and such that*

$$\int_{\mathbb{R}^n} \phi_\epsilon(|x|_2)\ dx = 1.$$

Define a function f_ϵ on $\mathbb{R}^n$ by

$$f_\epsilon(x) = \int_{\mathbb{R}^n} h_K(x + |x|_2 z) \phi_\epsilon(|z|_2)\ dz.$$

Then the function f_ϵ is infinitely differentiable on $\mathbb{R}^n \setminus \{0\}$ and is the support function of a convex body K_ϵ so that $f_\epsilon = h_{K_\epsilon}$ and for every $x \in S^{n-1}$,

$$|h_K(x) - h_{K_\epsilon}(x)| < C\epsilon.$$

Since the Minkowski functional of a convex body is the support function of the polar body, the latter theorem gives an approximation of any convex body by infinitely smooth convex bodies in the radial metric.

Also, considering bodies K_ϵ with Minkowski functionals

$$\|\cdot\|_{K_\epsilon} = \|\cdot\|_K + \epsilon|\cdot|_2,$$

one can approximate a convex body K in the radial metric by convex bodies with strictly positive curvature. More results on approximation with special properties can be found in [GrZ, Section 5].

An argument similar to that of Theorem 2.10 shows that one can approximate bodies in the norm of the space $C^k(S^{n-1})$ in the following sense: any k-smooth star body D can be approximated by a sequence of infinitely smooth star bodies D_m so that the radial functions ρ_{D_m} converge to ρ_D in the metric of the space $C^k(S^{n-1})$.

2.3. Radon transforms

The Radon transform and its connection with the Fourier transform play an important role in this text. We refer the reader to the books by Helgason [H1], [H2] for a systematic study of the Radon transforms.

Let ϕ be an integrable function on $\mathbb{R}^n$, which is also integrable on every hyperplane. The *Radon transform* of the function ϕ is defined as a function of $(\xi; t)$, $\xi \in S^{n-1}$, $t \in \mathbb{R}$:

$$\mathcal{R}\phi(\xi;t) = \int_{(x,\xi)=t} \phi(x)\, dx.$$

We frequently use a well-known connection between the Radon and Fourier transforms:

LEMMA 2.11. *For a fixed ξ, the Fourier transform of the function $g(t) = \mathcal{R}\phi(\xi;t)$, $t \in \mathbb{R}$, is equal to the function $z \mapsto \hat{\phi}(z\xi)$, $z \in \mathbb{R}$.*

PROOF. Making a change of variables $(x, \xi) = t$, we get

$$\hat{\phi}(z\xi) = \int_{\mathbb{R}^n} \phi(x) e^{-iz(x,\xi)}\, dx = \int_{\mathbb{R}} e^{-izt} \left(\int_{(x,\xi)=t} \phi(x)\, dx \right) dt = \hat{g}(z).$$

□

The *spherical Radon transform* $R : C(S^{n-1}) \mapsto C(S^{n-1})$ is a linear operator defined by

$$Rf(\xi) = \int_{S^{n-1}\cap\xi^\perp} f(x)\, dx, \quad \xi \in S^{n-1},$$

for every function $f \in C(S^{n-1})$. Using (2.2), one can express the volume of central hyperplane sections in terms of the spherical Radon transform: for every origin-symmetric star body K in $\mathbb{R}^n$ and every $\xi \in S^{n-1}$,

$$A_{K,\xi}(0) = \mathrm{Vol}_{n-1}(K \cap \xi^\perp) = \frac{1}{n-1} R(\|\cdot\|_K^{-n+1})(\xi). \tag{2.17}$$

Let $G(n,m)$ be the Grassman manifold of m-dimensional linear subspaces of $\mathbb{R}^n$, and let dH be the rotationally invariant probability measure on $G(n,m)$. The m-dimensional spherical Radon transform is an operator $R_m : C(S^{n-1}) \to C(G(n,m))$ defined by

$$R_m f(H) = \int_{S^{n-1}\cap H} f(x)\,dx.$$

The dual transform R_m^* (see [H2, p. 144] or [SW, pp. 146-150]) is an operator from $C(G(n,m))$ to $C(S^{n-1})$ such that, for every $F \in C(G(n,m))$,

$$R_m^* F(\xi) = \int_{\xi\in H} F(H)\,dH, \qquad \xi \in S^{n-1}, \tag{2.18}$$

with the latter integral being equal to

$$\int_{\xi\in H} F(H)\,dH = \frac{|S^{m-1}|}{|S^{n-1}|}\int_{G_\xi(n-1,m-1)} F(\mathrm{span}(H,\xi))\,dH, \tag{2.19}$$

where $G_\xi(n-1,m-1)$ stands for the Grassmanian of $(m-1)$-dimensional linear subspaces of the $(n-1)$-dimensional space $\xi^\perp$, and the rotationally invariant probability measure on this Grassmanian is still denoted by dH. Also $|S^{m-1}| = 2\pi^{m/2}/\Gamma(m/2)$ is the surface area of the unit sphere S^{m-1}.

The duality means that, for any functions $f \in C(S^{n-1})$ and $F \in C(G(n,m))$,

$$(R_m f, F) = \int_{G(n,m)} \left(\int_{S^{n-1}\cap H} f(\xi)\,d\xi \right) F(H)\,dH$$

$$= \int_{S^{n-1}} f(\xi) \left(\int_{\xi\in H} F(H)\,dH \right) d\xi = (f, R_m^* F). \tag{2.20}$$

We outline the proof of (2.20), communicated to us by S. Alesker. Let X be the set of pairs (ξ, H), where $\xi \in S^{n-1}$ and H is an m-dimensional subspace of $\mathbb{R}^n$ so that $\xi \in H$. The orthogonal group $O(n)$ acts on X transitively, so X admits a unique Haar probability measure μ. For any continuous function g on X, we have

$$\int_X g\,d\mu = c_1 \int_{G(n,m)} \left(\int_{\xi\in H} g(\xi,H)\,d\xi \right) dH \tag{2.21}$$

$$= c_2 \int_{S^{n-1}} \left(\int_{\xi\in H} g(\xi,H)\,dH \right) d\xi,$$

where c_1 and c_2 are some constants. In fact, all three integrals define continuous $O(n)$-invariant functionals on $C(X)$; therefore, all of them correspond to constant multiples of the Haar measure. To prove (2.20), consider any two functions $f \in C(S^{n-1})$ and $F \in C(G(n,m))$. Then put $g(\xi,H) = f(\xi)F(H)$ and compute the integrals in equation (2.21).

Put $F \equiv 1$ in equation (2.20). We get that, for any function $f \in C(S^{n-1})$,

$$\int_{G(n,m)} \left(\int_{S^{n-1}\cap H} f(\xi)\,d\xi \right) dH = \frac{|S^{m-1}|}{|S^{n-1}|} \int_{S^{n-1}} f(\xi)\,d\xi. \tag{2.22}$$

Note that, while the measure on the Grassmanian is probabilistic, the uniform measures on the spheres are not normalized in formulas (2.20) and (2.22), which explains the constant in the right-hand side of (2.22) (this constant can be verified by putting $f \equiv 1$).

The following lemma was proved in [K14, Lemma 1].

LEMMA 2.12. *Let $1 \le m < n$, and let $f, g \in C(S^{n-1})$. Then*

$$\int_{S^{n-1}} f(\xi) \left(\int_{S^{n-1}\cap\xi^\perp} g(x)\, dx \right) d\xi$$

$$= c(n,m) \int_{G(n,m)} \left(\int_{S^{n-1}\cap H} f(\xi) d\xi \right) \left(\int_{S^{n-1}\cap H^\perp} g(x) dx \right) dH,$$

where

$$c(n,m) = (|S^{n-1}||S^{n-2}|)/(|S^{m-1}||S^{n-m-1}|).$$

PROOF. Apply equality (2.20) to the functions f and

$$F(H) = \int_{S^{n-1}\cap H^\perp} g(x)\, dx.$$

We get

$$\int_{G(n,m)} \left(\int_{S^{n-1}\cap H} f(\xi) d\xi \right) \left(\int_{S^{n-1}\cap H^\perp} g(x) dx \right) dH$$

$$= c \int_{S^{n-1}} f(\xi) \left(\int_{\xi\in H} \left(\int_{S^{n-1}\cap H^\perp} g(x)\, dx \right) dH \right) d\xi, \tag{2.23}$$

with a constant to be computed later. Now apply (2.19) and change the variables $H \to H^\perp$:

$$\int_{\xi\in H} \left(\int_{S^{n-1}\cap H^\perp} g(x)\, dx \right) dH = c_1 \int_{G_\xi(n-1,n-m)} \left(\int_{S^{n-1}\cap H} g(x)\, dx \right) dH,$$

where $G_\xi(n-1, n-m)$ is the Grassmanian of $(n-m)$-dimensional subspaces of $\xi^\perp$ with its rotationally invariant probability measure. With (2.22) applied to the unit sphere $S^{n-1} \cap \xi^\perp$ of the hyperplane $\xi^\perp$, the latter is equal to

$$c_2 \int_{S^{n-1}\cap\xi^\perp} g(x)\, dx,$$

which proves the lemma up to a constant. To compute the constant, put $f \equiv 1$ and $g \equiv 1$. □

Applying Lemma 2.12 with $m = n-1$, we see that the spherical Radon transform $R = R_{n-1}$ is self-dual (see [Gr, Lemma 1.3.3] for a simple proof not involving Grassmanians).

LEMMA 2.13. *For any functions $f, g \in C(S^{n-1})$,*

$$\int_{S^{n-1}} Rf(\xi)\, g(\xi)\, d\xi = \int_{S^{n-1}} f(\xi)\, Rg(\xi)\, d\xi. \tag{2.24}$$

Finally, we extend the spherical Radon transform to measures. Let μ be a finite Borel measure on S^{n-1}. We define the spherical Radon transform of μ as a functional $R\mu$ on the space $C(S^{n-1})$ acting by

$$(R\mu, f) = (\mu, Rf) = \int_{S^{n-1}} Rf(x) d\mu(x).$$

By F. Riesz's characterization of continuous linear functionals on the space $C(S^{n-1})$ (see [DS, p. 262]) $R\mu$ is also a finite Borel measure on S^{n-1}. If μ has continuous density g, then, by Lemma 2.13, the Radon transform of μ has density Rg.

2.4. The Gamma-function

For $z \in \mathbb{C},\ \Re z > 0$, the Γ-function is defined by

$$\Gamma(z) = \int_0^\infty t^{z-1} e^{-t} dt. \tag{2.25}$$

A simple change of variables shows that, for any $p, q > 0$,

$$\int_0^\infty x^{p-1} e^{-x^q} dx = \frac{\Gamma(p/q)}{q}. \tag{2.26}$$

Integrating (2.25) by parts, we get that for $z \in \mathbb{C},\ \Re z > 0$,

$$\Gamma(z+1) = z\Gamma(z). \tag{2.27}$$

Since $\Gamma(1) = 1$, (2.27) implies $\Gamma(n+1) = n!$ for all $n \in \mathbb{N}$. Computing the integral

$$\int_0^\infty \int_0^\infty e^{-x^2-y^2} dx\, dy = \left(\int_0^\infty e^{-x^2} dx\right)^2$$

in polar coordinates and using (2.26), we get

$$\Gamma(1/2) = \sqrt{\pi}. \tag{2.28}$$

LEMMA 2.14. *The function* $\ln \Gamma(\cdot)$ *is convex on* $(0, \infty)$.

PROOF. By the Cauchy-Schwartz inequality, for $x_1, x_2 > 0$,

$$\Gamma\left(\frac{x_1+x_2}{2}\right) = \int_0^\infty t^{\frac{x_1+x_2}{2}-1} e^{-t} dt = \int_0^\infty t^{(x_1-1)/2} t^{(x_2-1)/2} e^{-t} dt$$

$$\leq \left(\int_0^\infty t^{x_1-1} e^{-t} dt\right)^{1/2} \left(\int_0^\infty t^{x_2-1} e^{-t} dt\right)^{1/2} = \sqrt{\Gamma(x_1)\Gamma(x_2)},$$

which implies the result. □

Differentiating under the integral in (2.25) by z, one can see that the Γ-function is analytic in $\{z \in \mathbb{C} : \Re z > 0\}$. We can use formula (2.27) to extend the Γ-function to an analytic function in $\mathbb{C} \setminus (-\mathbb{N} \cup \{0\})$ with simple poles at non-positive integers. The residue of the Γ-function at $z = -k$ is equal to $(-1)^k/k!$.

The Γ-function is positive on the intervals $(0, \infty)$ and $(-2k, -2k+1),\ k \in \mathbb{N}$, and it is negative on each of the intervals $(-2k+1, -2k+2),\ k \in \mathbb{N}$.

The analyticity of the Γ-function allows us to prove different formulas on an interval and then extend them to larger sets using the uniqueness of analytic continuation.

LEMMA 2.15. *If* x *and* $x + 1/2$ *are complex numbers that do not belong to the set* $-\mathbb{N} \cup \{0\}$, *then*

$$\Gamma(2x) = \frac{2^{2x-1}}{\sqrt{\pi}} \Gamma(x)\Gamma(x+\frac{1}{2}). \tag{2.29}$$

PROOF. We can prove (2.29) on any interval and then use analytic continuation. For $a, b > 0$, consider the double integral

$$\Gamma(a)\Gamma(b) = \int_0^\infty \int_0^\infty x^{a-1} y^{b-1} e^{-x-y} dx\, dy.$$

Substituting $x = u^2$, $y = v^2$, we write the latter integral as

$$4\int_0^\infty \int_0^\infty u^{2a-1}v^{2b-1}e^{-u^2-v^2}du\,dv.$$

Passing to polar coordinates and using (2.26), we get

$$2\Gamma(a+b)\int_0^{\pi/2}(\cos\theta)^{2a-1}(\sin\theta)^{2b-1}d\theta.$$

We get a formula for the Beta-function B defined by

$$B(a,b) = \frac{\Gamma(a)\Gamma(b)}{\Gamma(a+b)} = 2\int_0^{\pi/2}(\cos\theta)^{2a-1}(\sin\theta)^{2b-1}d\theta. \tag{2.30}$$

Now making a substitution $t = \cos\theta$,

$$B(a,b) = \int_{-1}^1 |t|^{2a-1}(1-t^2)^{b-1}dt. \tag{2.31}$$

Making another substitution $x = t^2$,

$$B(a,b) = \int_0^1 x^{a-1}(1-x)^{b-1}dx.$$

Putting $a = b$, we get

$$\begin{aligned} B(a,a) &= \int_0^1 \left(\frac{1}{4} - (\frac{1}{2} - x)^2\right)^{a-1} dx \\ &= 2\int_0^{1/2} \left(\frac{1}{4} - (\frac{1}{2} - x)^2\right)^{a-1} dx. \end{aligned}$$

After the additional substitution $1/2 - x = (1/2)\sqrt{u}$, we write the latter integral as

$$\frac{1}{2^{2a-1}}\int_0^1 u^{-1/2}(1-u)^{a-1}du = \frac{1}{2^{2a-1}}B(1/2,a).$$

Finally,

$$B(a,a) = \frac{1}{2^{2a-1}}B(1/2,a),$$

and the result follows from the definition of the Beta-function and (2.28). □

Next we compute the Euler integral:

LEMMA 2.16. *For* $0 < p < 1$,

$$\int_0^\infty \frac{x^{p-1}}{1+x^2}dx = \frac{\pi}{2\sin\frac{\pi p}{2}}.$$

PROOF. Consider the function $f(z) = z^{p-1}/(1+z^2)$, $z \in \mathbb{C}$. This function has simple poles at $z = i$ and $z = -i$. Take $0 < t < T$ and consider a contour starting at the point $(t, 0)$, going along the x-axis to the point $(T, 0)$, then around the circle with center at the origin and radius T, then along the line $ue^{2\pi i}, u \in [T, t]$, and, finally, around the circle with center at the origin and radius t. Integrals over the circles vanish, as $t \to 0$ and $T \to \infty$. We get the result by computing the residues of f at $z = \pm i$ and performing contour integration. □

LEMMA 2.17. *If* $x \in \mathbb{C}$ *is not a real integer, then*

$$\Gamma(x)\Gamma(1-x) = \frac{\pi}{\sin\pi x}. \tag{2.32}$$

PROOF. Put $a = 1 - b,\ b \in (0, 1/2)$, in (2.30). We get

$$\Gamma(b)\Gamma(1-b) = 2\int_0^{\pi/2} (\tan\theta)^{2b-1}\,d\theta.$$

Making the change of variables $x = \tan\theta$ and using Lemma 2.16, we get (2.32) for $x \in (0, 1/2)$. Then use analytic continuation. □

Our next formula will be used in Lemma 2.23.

LEMMA 2.18. *If $q \in \mathbb{C}$ is not a non-negative even integer and not a negative odd integer, then*

$$\frac{2^{q+1}\sqrt{\pi}\,\Gamma((q+1)/2)}{\Gamma(-q/2)} = -2\Gamma(1+q)\sin(\pi q/2). \tag{2.33}$$

PROOF. Write $\Gamma(1+q)$ using (2.29) with $x = (q+1)/2$, and then use (2.32) with $x = -q/2$. □

For $p > 0$, let B_p^n be the unit ball of the space ℓ_p^n; see (1.4).

LEMMA 2.19.

$$\mathrm{Vol}_n(B_p^n) = \frac{2^n\left(\Gamma(1+\frac{1}{p})\right)^n}{\Gamma(1+\frac{n}{p})}.$$

PROOF. Using (2.26) and (2.27), we compute the integral

$$\int_{\mathbb{R}^n} e^{-|x_1|^p - \ldots - |x_n|^p}\,dx = \left(\frac{2\Gamma(1/p)}{p}\right)^n = 2^n\left(\Gamma(1+\frac{1}{p})\right)^n.$$

On the other hand, compute the same integral in polar coordinates and use (2.26) and the polar formula for the volume (2.4):

$$= \int_{S^{n-1}}\left(\int_0^\infty r^{n-1}e^{-r^p\|\theta\|_p^p}\,dr\right)d\theta = \frac{\Gamma(n/p)}{p}\int_{S^{n-1}}\|\theta\|_p^{-n}\,d\theta$$

$$= n\mathrm{Vol}_n(B_p^n)\frac{\Gamma(n/p)}{p} = \mathrm{Vol}_n(B_p^n)\Gamma(1+\frac{n}{p}).$$

Comparing these two expressions for the same integral, we get the result. □

COROLLARY 2.20. *The volume and surface area of the unit Euclidean ball are equal to*

$$\mathrm{Vol}(B_2^n) = \frac{\pi^{n/2}}{\Gamma(1+\frac{n}{2})} \quad \text{and} \quad |S^{n-1}| = \frac{2\pi^{n/2}}{\Gamma(n/2)}.$$

PROOF. To get the formula for the surface area, write the integral

$$\mathrm{Vol}(B_2^n) = \int_{\mathbb{R}^n} \chi(|x|_2)\,dx$$

in polar coordinates, where, as before, χ is the indicator function of the interval $[-1, 1]$. □

2.5. The Fourier transform of distributions

The Fourier transform of distributions is the main tool used in this book. In this section we collect definitions, notation and results from the theory of distributions that we need later. We refer the reader to the books by Rudin [Ru] and Gelfand and Shilov [GS] for systematic studies of distributions. Note that in this text we use a definition of the Fourier transform which is slightly different from those in [Ru] and [GS]. We define the Fourier transform of a function $\phi \in L_1(\mathbb{R}^n)$ by

$$\mathcal{F}\phi(\xi) = \hat{\phi}(\xi) = \int_{\mathbb{R}^n} \phi(x)e^{-i(x,\xi)}dx, \quad \xi \in \mathbb{R}^n.$$

Rudin [Ru] uses the normalized Lebesgue measure, i.e. Lebesgue measure multiplied by the number $(2\pi)^{-n/2}$, while Gelfand and Shilov use Lebesgue measure without a coefficient, but they have $\exp(i(x,\xi))$ in place of $\exp(-i(x,\xi))$. The reason for our choice is that it is easy to adjust the results from [Ru], and, at the same time, we can directly use computations of the Fourier transform of certain distributions given in [GS], because the functions, for which we compute the Fourier transform in the sense of distributions, are always real-valued and even. For the same reason, we follow [Ru] by defining the action of a complex-valued function f on a test function ϕ as

$$\langle f, \phi \rangle = \int_{\mathbb{R}^n} f(x)\phi(x)\, dx,$$

while [GS] has this definition with the complex conjugate of $f(x)$. Finally, we again follow [Ru] in defining the Fourier transform of a distribution f by

$$\langle \hat{f}, \phi \rangle = \langle f, \hat{\phi} \rangle,$$

while the definition in [GS] is different:

$$\langle \hat{f}, \hat{\phi} \rangle = (2\pi)^n \langle f, \phi \rangle.$$

Let us stress once again that the definitions below are chosen in such a way that all the results from [Ru] can be used up to the coefficient $(2\pi)^{-n/2}$ in the formulas related to the Fourier transform, and all the results from [GS] can be used without any changes, because we always deal with real-valued even distributions (except for one occasion in Lemma 4.16), for which our formulas coincide with those from [GS].

Our choice of the space of test functions is the Schwartz space of rapidly decreasing infinitely differentiable functions. For a multiindex $\alpha = (\alpha_1, ..., \alpha_n)$, where $\alpha_k \in \mathbb{N} \cup \{0\}$, let $|\alpha| = \sum_{k=1}^n \alpha_k$. Let D^α be a differential operator defined for every $f \in C^{|\alpha|}(\mathbb{R}^n)$ by

$$D^\alpha f = \frac{\partial^{|\alpha|} f}{\partial x_1^{\alpha_1} ... \partial x_n^{\alpha_n}}.$$

We denote by $\mathcal{S} = \mathcal{S}(\mathbb{R}^n)$ the space of complex-valued functions $\phi \in C^\infty(\mathbb{R}^n)$ converging to zero at infinity together with all their derivatives faster than any negative power of $|\cdot|_2$, i.e. for every $k \in \mathbb{N}$,

$$p_k(\phi) = \sup_{|\alpha| \le k} \sup_{x \in \mathbb{R}^n} (1 + |x|_2)^k |D^\alpha \phi(x)| < \infty.$$

Throughout the text, elements of the space $\mathcal{S}$ will be called test functions.

A locally convex topology on $\mathcal{S}$ is generated by a sequence of seminorms p_k (see [Ru, Th 1.37]), where the local base of open sets is given by intersections of finite collections of sets of the form

$$V(k,m) = \{\phi \in \mathcal{S} : \; p_k(\phi) < \frac{1}{m}\}, \qquad k \in \mathbb{N}, \; m \in \mathbb{N}.$$

As usual, we denote by $\mathcal{S}'$ the space of linear continuous functionals on $\mathcal{S}$, which we call distributions over $\mathcal{S}$. Suppose that f is a locally integrable complex-valued function on $\mathbb{R}^n$ with *power growth at infinity*, i.e. there exists a number $\beta > 0$ so that

$$\lim_{|x|\to\infty} \frac{f(x)}{|x|_2^\beta} = 0.$$

Then f represents a distribution acting by integration: for every $\phi \in \mathcal{S}$,

$$\langle f, \phi \rangle = \int_{\mathbb{R}^n} f(x)\phi(x)\, dx.$$

For an open set Ω in $\mathbb{R}^n$, denote by $\mathcal{S}(\Omega)$ the space of test functions with support in Ω. Suppose that $f \in \mathcal{S}'$ and W is the union of all open sets Ω in $\mathbb{R}^n$ on which $f \equiv 0$, i.e. $\langle f, \phi\rangle = 0$ for every $\phi \in \mathcal{S}(\Omega)$. Then the set $\operatorname{supp}(f) = \mathbb{R}^n \setminus W$ is called the *support of the distribution* f. If $\phi \in \mathcal{S}$ is a test function so that $\operatorname{supp}(\phi) \cap \operatorname{supp}(f) = \emptyset$, then $\langle f, \phi \rangle = 0$.

An infinitely differentiable function g on $\mathbb{R}^n$ is called a *multiplicator* over $\mathcal{S}$ if $g\phi \in \mathcal{S}$ for every $\phi \in \mathcal{S}$ and the mapping $\phi \mapsto g\phi$ is continuous on $\mathcal{S}$. Every function $\phi \in \mathcal{S}$ is a multiplicator over $\mathcal{S}$, and so is every polynomial; see [Ru, Th. 7.4].

If ϕ is a test function, then so is its Fourier transform

$$\mathcal{F}\phi(x) = \hat{\phi}(x) = \int_{\mathbb{R}^n} \phi(\xi) e^{-i(x,\xi)}\, d\xi;$$

see [Ru, Th. 7.4]. The Fourier transform is invertible on $\mathcal{S}$, and the inverse operator is given by

$$\tilde{\mathcal{F}}\phi(\xi) = (2\pi)^{-n} \int_{\mathbb{R}^n} \phi(x) e^{i(x,\xi)} dx,$$

see [Ru, Th 7.7]. It follows that, for every $\phi \in \mathcal{S}$,

$$(\hat{\phi})^\wedge(\xi) = (2\pi)^n \phi(-\xi).$$

Also both operators $\mathcal{F}$ and $\tilde{\mathcal{F}}$ are continuous on the space $\mathcal{S}$, [Ru, Th. 7.7].

An easy application of the Fubini theorem is the following *Parseval formula* for test functions. If ϕ, ψ are test functions, then

$$\int_{\mathbb{R}^n} \hat{\phi}(x)\, \psi(x)\, dx = \int_{\mathbb{R}^n} \phi(\xi)\, \hat{\psi}(\xi)\, d\xi. \tag{2.34}$$

In fact, the left-hand side can be written as

$$\int_{\mathbb{R}^n} \left(\int_{\mathbb{R}^n} \phi(\xi) e^{-i(x,\xi)}\, d\xi \right) \psi(x)\, dx,$$

and, by the Fubini theorem, the latter is equal to

$$\int_{\mathbb{R}^n} \left(\int_{\mathbb{R}^n} \psi(x) e^{-i(x,\xi)}\, dx \right) \phi(\xi)\, d\xi = \int_{\mathbb{R}^n} \phi(\xi)\, \hat{\psi}(\xi)\, d\xi.$$

Since test functions are dense in the spaces $L_1(\mathbb{R}^n)$ and $L_2(\mathbb{R}^n)$, Parseval's formula can be extended to the cases where both functions ϕ and ψ belong to $L_1(\mathbb{R}^n)$ or both belong to $L_2(\mathbb{R}^n)$ (see [Ru, Th 7.9]).

Parseval's formula motivates the definition of the *Fourier transform of a distribution* f as a distribution $\hat{f}$ acting by

$$\langle \hat{f}, \phi \rangle = \langle f, \hat{\phi} \rangle$$

for every test function ϕ. Since the mapping $\phi \mapsto \hat{\phi}$ is an isomorphism of $\mathcal{S}(\mathbb{R}^n)$, the Fourier transform of a distribution is well defined, and $\hat{f} = \hat{g}$ implies $f = g$. When g is given by a longer expression, we write $(g)^\wedge$ for the Fourier transform of g.

If a test function ϕ is even, we have

$$(\hat{\phi})^\wedge = (2\pi)^n \phi \qquad \text{and} \qquad \langle \hat{f}, \hat{\phi} \rangle = (2\pi)^n \langle f, \phi \rangle$$

for every $f \in \mathcal{S}'$. Throughout this text (except for one occasion in Lemma 4.16) we consider only real-valued even test functions ϕ, for which the Fourier transform $\hat{\phi}$ is also an even real-valued function.

A distribution f is called even homogeneous of degree $p \in \mathbb{R}$ if

$$\langle f(x), \phi(x/\alpha) \rangle = |\alpha|^{n+p} \langle f, \phi \rangle \tag{2.35}$$

for every test function ϕ and every $\alpha \in \mathbb{R}$, $\alpha \neq 0$.

LEMMA 2.21. *The Fourier transform of an even homogeneous distribution of degree p is an even homogeneous distribution of degree $-n-p$.*

PROOF. A simple computation shows that, for any test function ϕ and any $\alpha \neq 0$,

$$\widehat{\phi(x/\alpha)}(\xi) = |\alpha|^n \hat{\phi}(\alpha \xi).$$

By (2.35),

$$\begin{aligned} \langle \hat{f}(x), \phi(x/\alpha) \rangle &= \langle f, \widehat{\phi(x/\alpha)} \rangle = \langle f(\xi), |\alpha|^n \hat{\phi}(\alpha\xi) \rangle \\ &= |\alpha|^{-p} \langle f, \hat{\phi} \rangle = |\alpha|^{n+(-n-p)} \langle \hat{f}, \phi \rangle, \end{aligned}$$

which means that $\hat{f}$ is an even homogeneous distribution of degree $-n-p$. □

For any multiindex α, the derivative of the order α of a distribution f is defined by

$$\langle D^\alpha f, \phi \rangle = (-1)^{|\alpha|} \langle f, D^\alpha \phi \rangle.$$

The Fourier transform is related to differentiation as follows:

$$(D^\alpha f)^\wedge = i^{|\alpha|} x_1^{\alpha_1} \dots x_n^{\alpha_n} \hat{f}, \tag{2.36}$$

and

$$D^\alpha \hat{f} = (-i)^{|\alpha|} (x_1^{\alpha_1} \dots x_n^{\alpha_n} f(x))^\wedge;$$

see [Ru, Th 7.4]. The product of a polynomial and a distribution is defined by

$$\langle x_1^{\alpha_1} \dots x_n^{\alpha_n} f, \phi \rangle = \langle f, x_1^{\alpha_1} \dots x_n^{\alpha_n} \phi \rangle,$$

which is possible because every polynomial is a multiplicator. The connection between the Fourier transform and differentiation follows from similar properties of test functions, which can be proved by simple integration by parts.

If a distribution is supported in the origin, then it is a finite linear combination of derivatives of the δ-function defined by $\langle \delta, \phi \rangle = \phi(0)$, [Ru, Th. 6.25]. Since the Fourier transform of a linear combination of derivatives of the δ-function is a polynomial, any two distributions, whose Fourier transforms are equal outside of the origin, can differ only by a polynomial; see [Ru, Section 7.16].

For any open set $Q \subset \mathbb{C}$ we say that a family of distributions f_q, $q \in Q$, is analytic on Q if for every test function ϕ, the function

$$q \mapsto \langle f_q, \phi \rangle$$

is analytic on Q. This concept is important for us, because we usually prove certain formulas for $q \in (-1, 0)$ and then use analytic continuation to prove these formulas for other values of q. We often use the following fact which can be found, in a similar form, in [GS, Ch. 1, Sect. 3]:

LEMMA 2.22. *For any star body K in $\mathbb{R}^n$ the family of distributions $\|\cdot\|_K^q$ is analytic in the domain $q \in \mathbb{C}$, $\Re(q) > -n$. The family of Fourier transforms $(\|\cdot\|_K^q)^\wedge$ is also analytic in the same domain.*

PROOF. By Lemma 2.1, the function $\|\cdot\|_K^q$ is locally integrable on $\mathbb{R}^n$. Therefore, for any test function ϕ,

$$h(q) = \langle \|\cdot\|_K^q, \phi \rangle = \int_{\mathbb{R}^n} \|x\|_K^q \phi(x)\, dx.$$

Differentiating under the integral, it is easy to see that the derivative $h'(q)$ exists everywhere in the domain $\Re(q) > -n$ and is equal to

$$h'(q) = \int_{\mathbb{R}^n} \|x\|_K^q \ln(\|x\|_K)\phi(x)\, dx.$$

Note that an easy way of proving the possibility of differentiation by q under the integral is to do it in the opposite direction – integrate by q the equality for the derivatives and use the Fubini theorem.

The second statement follows from the first and the definition of the Fourier transform, because $\langle (\|\cdot\|_K)^\wedge, \phi \rangle = \langle \|\cdot\|_K^q, \hat{\phi} \rangle$. □

A crucial role in what follows belongs to the distributions $t \mapsto |t|^q$, $t \in \mathbb{R}$, where $q \in \mathbb{C}$. If $\Re(q) \leq -1$, this function is not locally integrable on $\mathbb{R}$, so to define the corresponding distribution, we need a special procedure that is called *regularization*. We want this procedure to produce an analytic family of distributions on $\mathbb{C}$ without negative integers.

For $t \in \mathbb{R}$, let $t_+ = \max\{0, t\}$. We first define the distributions t_+^q. If $\Re q > -1$, then the function t_+^q is locally integrable and its action on a test function ϕ is defined by

$$\langle t_+^q, \phi(t) \rangle = \int_0^\infty t^q \phi(t)\, dt.$$

We write the right-hand side of the latter equality in a different form:

$$\int_0^\infty t^q \phi(t)\, dt = \int_0^1 t^q(\phi(t) - \phi(0))\, dt + \int_1^\infty t^q \phi(t) dt + \frac{\phi(0)}{1+q}.$$

Now the integrals in the right-hand side converge absolutely for every q with $\Re q \in (-2, -1)$, and we define the distributions t_+^q with $\Re q \in (-2, -1)$ by

$$\langle t_+^q, \phi(t) \rangle = \int_0^1 t^q(\phi(t) - \phi(0))\, dt + \int_1^\infty t^q \phi(t) dt + \frac{\phi(0)}{1+q}. \tag{2.37}$$

In the case $\Re q > -1$ the latter equality coincides with the original definition, so we can say that formula (2.37) defines the distributions t_+^q for all $q \in \mathbb{C}$, $\Re q > -2$, $q \neq -1$. Moreover, by the same argument as in Lemma 2.22, the family of distributions $q \mapsto t_+^q$ is analytic in this domain.

We now continue this procedure: for $q \in \mathbb{C}$ and $m \in \mathbb{N}$ such that $-m-1 < \Re q$, $q \neq -1, -2, \ldots, -m$, and for every $\phi \in \mathcal{S}$, we put

$$\langle t_+^q, \phi(t)\rangle = \int_0^1 t^q \left(\phi(t) - \phi(0) - t\phi'(0) - \cdots - \frac{t^{m-1}}{(m-1)!}\phi^{(m-1)}(0)\right) dt$$

$$+ \int_1^\infty t^q \phi(t) dt + \sum_{k=1}^m \frac{\phi^{(k-1)}(0)}{(k-1)!(q+k)}. \tag{2.38}$$

If $-m-1 < \Re(q) < -m$, we have

$$\langle t_+^q, \phi(t)\rangle = \int_0^\infty t^q \left(\phi(t) - \phi(0) - t\phi'(0) - \cdots - \frac{t^{m-1}}{(m-1)!}\phi^{(m-1)}(0)\right) dt,$$

since in this case

$$\int_1^\infty t^{q+k-1} dt = -\frac{1}{q+k},$$

for $k = 1, \ldots, m$. It can be checked again by direct differentiation that the family

$$\{t_+^q : q \in \mathbb{C} \setminus \{-1, -2, \ldots\}\}$$

forms an analytic distribution, that is, for any $\phi \in \mathcal{S}$, the function $q \mapsto \langle t_+^q, \phi(t)\rangle$ is analytic on $\mathbb{C} \setminus \{-1, -2, \ldots\}$.

It is clear that $\langle t_+^q, \phi(t)\rangle$ has, for each $k \in \mathbb{N}$, a simple pole at $q = -k$ with residue $\phi^{(k-1)}(0)/(k-1)!$. The function $q \mapsto \Gamma(q+1)$ also has a simple pole at $q = -k$ with residue $(-1)^{k-1}/(k-1)!$; see Section 2.4. We conclude that we can extend

$$\left\{\frac{t_+^q}{\Gamma(q+1)} : q \in \mathbb{C} \setminus \{-1, -2, \ldots\}\right\}$$

to an analytic distribution on $\mathbb{C}$, still denoted by $\{t_+^q/\Gamma(q+1) : q \in \mathbb{C}\}$, and for $q = -k$ and $\phi \in \mathcal{S}$,

$$\left\langle \frac{t_+^q}{\Gamma(q+1)}, \phi(t)\right\rangle = (-1)^{k-1}\phi^{(k-1)}(0).$$

Outside of any neighborhood of 0, the functional $t_+^q/\Gamma(q+1)$ acts like a continuous function, so we can actually apply $t_+^q/\Gamma(q+1)$ to any continuous function that is infinitely differentiable in a neighborhood of 0, drawing the same conclusions as for functions in $\mathcal{S}$. We are going to do it later in the definition of fractional derivatives.

For $q \in \mathbb{C} \setminus -2\mathbb{N}+1$, we define the distribution $|t|^q$ by

$$\langle |t|^q, \phi\rangle = \langle t_+^q, \phi\rangle + \langle t_+^q, \phi(-t)\rangle.$$

For any $\phi \in \mathcal{S}$, the function $q \mapsto \langle |t|^q, \phi\rangle$ is analytic in $\mathbb{C} \setminus -2\mathbb{N}+1$. Note that negative even integers are included, because the singularities in (2.38) corresponding to even k disappear after adding the same expression with $\phi(-t)$.

We now compute the Fourier transform of $|t|^q$. We present this computation with all the details, some of which will be omitted in similar calculations later.

LEMMA 2.23. *If $q \in \mathbb{C}$ is not a non-negative even integer and not a negative odd integer, then the Fourier transform of the distribution $|t|^q$, $t \in \mathbb{R}$, is equal to*

$$(|t|^q)^\wedge(\xi) = -2\Gamma(1+q)\sin(\pi q/2)\,|\xi|^{-q-1}, \quad \xi \in \mathbb{R}.$$

PROOF. First, suppose that $q \in (-1,0)$. Making the substitution $u = zt$ in the integral and using (2.26), we get

$$|t|^q = \frac{2^{q/2+1}}{\Gamma(-q/2)} \int_0^\infty z^{-1-q} e^{-z^2t^2/2}\, dz.$$

We often use representations like this to compute the Fourier transform. The reason is that the Fourier transform of the Gaussian density is well known (and easily computable):

$$(e^{-t^2/2})^\wedge(z) = \sqrt{2\pi} e^{-z^2/2}.$$

Now for any even test function ϕ :

$$\langle (|t|^q)^\wedge, \phi \rangle = \langle |t|^q, \hat{\phi} \rangle = \int_{\mathbb{R}} |t|^q \hat{\phi}(t)\, dt$$

$$= \frac{2^{q/2+1}}{\Gamma(-q/2)} \int_{\mathbb{R}} \left(\int_0^\infty z^{-1-q} e^{-z^2t^2/2}\, dz \right) \hat{\phi}(t)\, dt$$

$$= \frac{2^{q/2+1}}{\Gamma(-q/2)} \int_0^\infty z^{-1-q} \left(\int_{\mathbb{R}} e^{-(tz)^2/2} \hat{\phi}(t)\, dt \right) dz.$$

Applying Parseval's equality (2.34) to the integral over $\mathbb{R}$, the fact that

$$(f(tx))^\wedge(\xi) = t^{-1}\hat{f}(\xi/t) \tag{2.39}$$

and formula (2.26) (make the substitution $u = \xi/z$), we continue the calculation:

$$= \frac{2^{q/2+1}}{\Gamma(-q/2)} \int_0^\infty z^{-1-q} \left(\int_{\mathbb{R}} \sqrt{2\pi} z^{-1} e^{-\xi^2/2z^2} \phi(\xi)\, d\xi \right) dz$$

$$= \frac{\sqrt{2\pi}2^{q/2+1}}{\Gamma(-q/2)} \int_{\mathbb{R}} \left(\int_0^\infty z^{-2-q} e^{-\xi^2/2z^2} dz \right) \phi(\xi)\, d\xi$$

$$= \frac{2^{q+1}\sqrt{\pi}\Gamma((q+1)/2)}{\Gamma(-q/2)} \int_{\mathbb{R}} |\xi|^{-1-q} \phi(\xi)\, d\xi.$$

We see that for every even test function ϕ,

$$\langle (|t|^q)^\wedge, \phi \rangle = \frac{2^{q+1}\sqrt{\pi}\Gamma((q+1)/2)}{\Gamma(-q/2)} \langle |\xi|^{-1-q}, \phi \rangle. \tag{2.40}$$

Now we can use formula (2.33) to simplify the constant, which proves the lemma for every $q \in (-1,0)$. The distributions $|t|^q$ are defined in such a way that both sides of (2.40) are analytic functions of q everywhere in $\mathbb{C}$ without negative odd integers and non-negative even integers. By uniqueness of analytic continuation, we can extend the formula from the interval $(-1,0)$ to all such q. □

2.6. Fractional derivatives

Fractional derivatives play an important role in the Fourier analytic solution of the Busemann-Petty problem in [GKS]. However, we do not use fractional derivatives in full generality; we need fractional derivatives at only one point, the origin. In general, the fractional derivative of the order q of a test function $\phi \in \mathcal{S}(\mathbb{R})$ is defined as the convolution of ϕ with $t_+^{-1-q}/\Gamma(-q)$:

$$\phi^{(q)}(x) = \langle \frac{t_+^{-1-q}}{\Gamma(-q)}, \phi(x-t)\rangle.$$

We need this definition only with $x = 0$. This allows us to replace the test function ϕ by any function differentiable up to a certain order in a neighborhood of zero. Let us formulate this precisely, though the procedure is essentially the same as in the definition of the distributions t_+^q; see (2.38).

Let $m \in \mathbb{N} \cup \{0\}$ and suppose h is a continuous integrable function on $\mathbb{R}$ that is m times continuously differentiable in some neighborhood of zero.

For $q \in \mathbb{C}$, $-1 < \Re(q) < m$, $q \neq 0, 1, ..., m-1$, the *fractional derivative* of the order q of the function h at zero is defined as the action of the distribution $t_+^{-1-q}/\Gamma(-q)$ on the function h, as follows:

$$h^{(q)}(0) = \frac{1}{\Gamma(-q)} \int_0^1 t^{-1-q} \left(h(t) - h(0) - ... - h^{(m-1)}(0)\frac{t^{m-1}}{(m-1)!} \right) dt$$

$$+\frac{1}{\Gamma(-q)} \int_1^\infty t^{-1-q} h(t) dt + \frac{1}{\Gamma(-q)} \sum_{k=0}^{m-1} \frac{h^{(k)}(0)}{k!(k-q)}. \tag{2.41}$$

It is easy to see that for a fixed q the definition does not depend on the choice of $m > \Re(q)$, as long as f is m times continuously differentiable. Note that without dividing by $\Gamma(-q)$ the expression for the fractional derivative represents an analytic function in the domain $\{q \in \mathbb{C} : \Re(q) > -1\}$ not including integers and has simple poles at integers. The function $\Gamma(-q)$ is analytic in the same domain and also has simple poles at non-negative integers, so after the division we get an analytic function in the whole domain $\{q \in \mathbb{C} : m > \Re(q) > -1\}$, which also defines fractional derivatives of integer orders. Moreover, computing the limit as $q \to k$, where k is a non-negative integer, we see that the fractional derivatives of integer orders coincide with usual derivatives up to a sign (when we compute the limit, the first two summands in the right-hand side of (2.41) converge to zero, since $\Gamma(-q) \to \infty$, and the limit in the third summand can be computed using the property $\Gamma(x+1) = x\Gamma(x)$ of the Γ-function):

$$h^{(k)}(0) = (-1)^k \frac{d^k}{dt^k} h(t)|_{t=0}.$$

If h is an even function, its derivatives of odd orders at the origin are equal to zero and, for $m-2 < \Re(q) < m$, expression (2.41) becomes

$$h^{(q)}(0) = \frac{1}{\Gamma(-q)} \int_0^\infty t^{-q-1} \left(h(t) - \sum_{j=0}^{(m-2)/2} \frac{t^{2j}}{(2j)!} h^{(2j)}(0) \right) dt. \tag{2.42}$$

We also note that if $-1 < q < 0$, then

$$h^{(q)}(0) = \frac{1}{\Gamma(-q)} \int_0^\infty t^{-1-q} h(t)\, dt. \tag{2.43}$$

2.7. Positive definite distributions

We say that a distribution f is *positive definite* if its Fourier transform is a *positive distribution*, i.e. we have $\langle \hat{f}, \phi \rangle \geq 0$ for every non-negative test function ϕ. L. Schwartz's generalization of Bochner's theorem (see, for example, [GV, p. 152]) states that a distribution is positive definite if and only if it is the Fourier transform of a tempered measure on $\mathbb{R}^n$. Recall that a (non-negative, not necessarily finite) measure μ is called tempered if

$$\int_{\mathbb{R}^n} (1 + |x|_2)^{-\beta}\, d\mu(x) < \infty$$

for some $\beta > 0$. We do not prove Schwartz's theorem here and instead refer the reader to [GV, p. 152]. However, we are going to prove a similar fact for distributions that are positive outside of the origin. This proof follows the steps of the proof of Schwartz's theorem.

We mostly deal with positive definite distributions that are at the same time homogeneous. The corresponding tempered measures are also homogeneous as distributions and can be written in polar coordinates, as follows:

LEMMA 2.24. *Let μ be a tempered measure on $\mathbb{R}^n$ which is, at the same time, an even homogeneous distribution of degree $-n+p$, $p \in (0, n)$. Then there exists a finite Borel measure μ_0 on the sphere S^{n-1} so that, for every even test function ϕ,*

$$\langle \mu, \phi \rangle = \int_{S^{n-1}} \left(\int_0^\infty t^{p-1} \phi(t\xi)\, dt \right) d\mu_0(\xi). \tag{2.44}$$

PROOF. Let us first show that μ cannot have an atom at the origin. In fact, suppose that $\mu = \mu_1 + a\delta$, where $\mu_1(\{0\}) = 0$ and δ is the unit mass at the origin. Since μ is homogeneous of degree $-n+p$, for every non-negative test function ϕ with $\phi(0) > 0$ and every $t > 0$, we have

$$\langle \mu, \phi(x/t) \rangle = t^p \langle \mu, \phi \rangle \to 0$$

as $t \to 0$. On the other hand,

$$\langle \mu, \phi(x/t) \rangle = \langle \mu_1, \phi(x/t) \rangle + a\phi(0) \geq a\phi(0),$$

so $a = 0$.

For every Borel subset $A \subset S^{n-1}$ and every interval $(a, b] \in [0, \infty)$ let

$$A \times (a, b] = \{x \in \mathbb{R}^n : \ x = t\theta,\ t \in (a, b],\ \theta \in A\},$$

and let $\chi_{A \times (a,b]}$ be the indicator of this set.

By the definition of a homogeneous distribution, we have

$$\int_{\mathbb{R}^n} \phi(x/t)\, d\mu(x) = \langle \mu, \phi(x/t) \rangle$$

$$= t^p \langle \mu, \phi \rangle = t^p \int_{\mathbb{R}^n} \phi(x)\, d\mu(x)$$

for every test function ϕ and $t > 0$. We approximate the function $\chi(A \times [0,1])$ by test functions and use the dominated convergence theorem to get that, for every open set $A \subset S^{n-1}$ and $t > 0$,

$$\mu(A \times [0,t]) = t^p \mu(A \times [0,1]).$$

Now, for every Borel subset $A \subset S^{n-1}$ and every $0 \le a < b$ we have

$$\mu(A \times (a,b]) = (b^p - a^p)\mu(A \times [0,1]).$$

Define a measure μ_0 on S^{n-1} by $\mu_0(A) = p\mu(A \times [0,1])$ for every Borel set $A \subset S^{n-1}$. Clearly,

$$\int_{S^{n-1}} \left(\int_{\mathbb{R}} |t|^{p-1} \chi_{A\times(a,b]}(t\theta)\, dt \right) d\mu_0(\theta) = \frac{b^p - a^p}{p} \mu_0(A) = \mu(\chi_{A\times(a,b]}).$$

Therefore, we get equality (2.44) with $\phi = \chi_{A\times(a,b]}$ and the result follows by approximation, since A, a, b are arbitrary. □

We often consider measures that are homogeneous distributions of degree $-n-p$, where $p > 0$. In this case the integral in (2.44) might diverge, so we consider only test functions ϕ supported outside of the origin.

Let D be an open set in $\mathbb{R}^n$, and let $K(D)$ be the space of test functions with compact support in D. We say that two distributions f, g are equal on D if $\langle f, \phi \rangle = \langle g, \phi \rangle$ for every $\phi \in K(D)$.

A distribution f is said to be positive (negative) in D if $\langle f, \phi \rangle \ge 0$ (≤ 0) for every non-negative test function $\phi \in K(D)$.

A sequence of non-negative test functions ω_k is called a *δ-sequence* if for each k the function ω_k is supported in the Euclidean ball with center at the origin and radius $1/k$ and if $\int_{\mathbb{R}^n} \omega_k = 1$. It is easy to see that for any continuous function h with compact support A in $\mathbb{R}^n$, the convolutions $h * \omega_k$ are test functions that converge to h in the uniform metric, as $k \to \infty$. For any open set D containing A, these convolutions are supported in D starting from some k. An example of a δ-sequence is

$$\omega_k(x) = c_k \exp\left(-\frac{1}{1 - k^2|x|_2^2} \right) \quad \text{if} \quad |x|_2 < \frac{1}{k},$$

and $\omega_k(x) = 0$ if $|x|_2 \ge 1/k$, where c_k is a constant so that $\int_{\mathbb{R}^n} \omega_k = 1$.

LEMMA 2.25. *Let $0 < a < b < \infty$. Suppose that F is a positive distribution in the domain $(a,b) \times S^{n-1}$. Then for every interval $[c,d] \subset (a,b)$ there exists a finite Borel measure μ on $[c,d] \times S^{n-1}$ so that $F = \mu$ on $[c,d] \times S^{n-1}$.*

If $p > 0$ and $F \in \mathcal{S}(\mathbb{R}^n)$ is a positive outside of the origin even homogeneous distribution of degree $-n-p$, then there exists a finite Borel measure μ_0 on S^{n-1} so that for every even test function ϕ with compact support outside of the origin,

$$\langle F, \phi \rangle = \int_{S^{n-1}} \left(\int_0^\infty t^{-1-p} \phi(t\xi) dt \right) d\mu_0(\xi).$$

PROOF. Let $\varepsilon > 0$ be such that $[c-\varepsilon, d+\varepsilon] \subset (a,b)$. There exists a test function ψ supported in $(a,b) \times S^{n-1}$ so that $\psi = 1$ everywhere on $[c-\varepsilon, d+\varepsilon] \times S^{n-1}$.

First, we show that F is continuous on $K([c-\varepsilon, d+\varepsilon] \times S^{n-1})$ in the topology of uniform convergence. Consider any sequence of test functions ϕ_k supported in $[c-\varepsilon, d+\varepsilon] \times S^{n-1}$ so that $\phi_k \to 0$ uniformly. This means that for every $\delta > 0$, starting from some k, $-\delta\psi < \phi_k < \delta\psi$ on $[c-\varepsilon, d+\varepsilon] \times S^{n-1}$. Since F is a positive distribution, starting from the same k, we have

$$-\delta\langle F, \psi\rangle \leq \langle F, \phi_k\rangle \leq \delta\langle F, \psi\rangle,$$

so the sequence $\langle F, \phi_k\rangle$ converges to zero, which establishes our claim.

Now we prove that F can be extended to a continuous functional on the space $C([c,d] \times S^{n-1})$. Let h be any continuous function on $[c,d] \times S^{n-1}$. By the remark before this lemma, we can construct a sequence ϕ_k of test functions supported in $[c-\varepsilon, d+\varepsilon] \times S^{n-1}$ that converges to h uniformly. This sequence is a Cauchy sequence in the uniform metric, so for every $\delta > 0$ we have

$$-\delta\psi \leq \phi_j - \phi_k \leq \delta\psi$$

for large enough j and k. Since F is a positive distribution, we have

$$-\delta\langle F, \psi\rangle \leq \langle F, \phi_j\rangle - \langle F, \phi_k\rangle \leq \delta\langle F, \psi\rangle,$$

so $\langle F, \phi_k\rangle$ is a Cauchy sequence. This sequence converges to a number that we call $F_e(h)$. By continuity of F on test functions with the topology of uniform convergence, the number $F_e(h)$ does not depend on the choice of a sequence ϕ_k, and F_e is a continuous linear functional on $C([c,d] \times S^{n-1})$. By F. Riesz's characterization of continuous linear functionals on the spaces $C(K)$ (see [DS, p.262]) this functional is a finite Borel measure on $[c,d] \times S^{n-1}$. Clearly, the distribution F coincides with μ on test functions supported in $[c,d] \times S^{n-1}$.

The proof of the second statement is similar to that of Lemma 2.24. By the first part of Lemma 2.25, the distribution F is equal to some finite Borel measure μ on $[1,2] \times S^{n-1}$. By homogeneity, using

$$\langle \mu, \phi(x/t)\rangle = |t|^{-p}\langle \mu, \phi\rangle,$$

we can extend this measure to $[2,4] \times S^{n-1}$ and so on, so that we get a Borel measure on $\mathbb{R}^n \setminus \{0\}$ that coincides with F on test functions with compact support outside of the origin. Then, approximating by test functions, we get that for any $0 < a < b < \infty$, any $k > 0$ and any Borel subset A of the sphere,

$$\mu([ka, kb) \times A) = k^{-p}\mu([a,b) \times A),$$

so

$$\mu([1,\infty) \times A) = \sum_{j=1}^{\infty} \mu([2^{j-1}, 2^j) \times A) = \mu([1,2) \times A) \sum_{j=1}^{\infty} 2^{-(j-1)p} < \infty,$$

and also

$$\mu([a,b) \times A) = (a^{-p} - b^{-p})\mu([1,\infty) \times A).$$

We can now define a measure μ_0 on the sphere by

$$\mu_0(A) = p\mu([1,\infty) \times A).$$

We get

$$\mu([a,b) \times A) = \int_{S^{n-1}} \left(\int_0^\infty t^{-1-p}\chi_{[a,b)\times A}(t\theta)dt \right) d\mu_0(\theta).$$

The result follows, since a, b, A are arbitrary. □

We are going to use Lemmas 2.24 and 2.25 in the following form. Note that the constant $\Gamma(-p/2)$ in the second part does not play any role here and is needed only for future applications.

COROLLARY 2.26. *(i) Let K be an origin-symmetric star body in $\mathbb{R}^n$ and let $p \in (0,n)$. The function $\|\cdot\|_K^{-p}$ represents a positive definite distribution on $\mathbb{R}^n$ if and only if there exists a finite Borel measure μ_0 on S^{n-1} so that for every even test function ϕ,*

$$\int_{\mathbb{R}^n} \|x\|_K^{-p}\phi(x)\, dx = \int_{S^{n-1}} \left(\int_0^\infty t^{p-1}\hat{\phi}(t\xi)dt\right) d\mu_0(\xi). \tag{2.45}$$

(ii) Suppose that $p > 0$ is not an even integer. The distribution $\Gamma(-p/2)(\|\cdot\|_K^p)^\wedge$ is positive on $\mathbb{R}^n \setminus \{0\}$ if and only if there exists a finite Borel measure μ_0 on S^{n-1} so that, for every even test function ϕ whose Fourier transform is supported in $\mathbb{R}^n \setminus \{0\}$,

$$\int_{\mathbb{R}^n} \|x\|_K^p\phi(x)dx = \frac{1}{\Gamma(-p/2)}\int_{S^{n-1}} \left(\int_0^\infty t^{-1-p}\hat{\phi}(t\xi)dt\right) d\mu_0(\xi). \tag{2.46}$$

PROOF. The "if" parts in both statements are straightforward. In fact, if we have (2.45), then for any non-negative even test function ϕ,

$$\langle(\|\cdot\|^{-p})^\wedge, \phi\rangle = \langle\|\cdot\|^{-p}, \hat{\phi}\rangle = \int_{\mathbb{R}^n} \|x\|^{-p}\hat{\phi}(x)\, dx$$

$$= (2\pi)^n \int_{S^{n-1}} \left(\int_0^\infty t^{p-1}\phi(t\xi)dt\right) d\mu_0(\xi) \geq 0.$$

The second statement is similar (note that we start with a non-negative test function ϕ supported in $\mathbb{R}^n \setminus \{0\}$ and then apply (2.46) to the function $\hat{\phi}$ in place of ϕ).

Now we prove the necessity. For the first part, by Schwartz's theorem, the positive definite distribution $\|\cdot\|_K^{-p}$ is the Fourier transform of a tempered measure μ which, by Lemma 2.21, is an even homogeneous distribution of degree $-n+p$. By Lemma 2.24, there exists a finite Borel measure μ_0 on S^{n-1} so that for any even test function ϕ,

$$\int_{\mathbb{R}^n} \|x\|_K^{-p}\phi(x)dx = \langle\hat{\mu}, \phi\rangle$$

$$= \langle\mu, \hat{\phi}\rangle = \int_{S^{n-1}} \left(\int_0^\infty t^{p-1}\hat{\phi}(t\xi)dt\right) d\mu_0(\xi).$$

For the second part, the function $\Gamma(-p/2)\|\cdot\|_K^p$ is the Fourier transform of the distribution

$$F = \frac{\Gamma(-p/2)}{(2\pi)^n}(\|\cdot\|_K^p)^\wedge,$$

which is positive in $\mathbb{R}^n \setminus \{0\}$ and, again by Lemma 2.21, even homogeneous of degree $-n-p$. For any even test function ϕ whose Fourier transform is supported outside of the origin, we have

$$\int_{\mathbb{R}^n} \|x\|_K^p\phi(x)dx = \frac{1}{\Gamma(-p/2)}\langle F, \hat{\phi}\rangle,$$

and the result follows from Lemma 2.25. $\square$

2.8. Stable random variables and the function γ_q

In this section we study a special function that will be used many times later. Denote by γ_q the Fourier transform of the function $z \mapsto \exp(-|z|^q)$, $z \in \mathbb{R}$. The functions γ_q can be computed precisely only for $q = 2$, where

$$\gamma_2(t) = (e^{-z^2})^\wedge(t) = \sqrt{\pi}\,\exp(-t^2/4),$$

and for $q = 1$, where

$$\gamma_1(t) = \frac{2}{1+t^2}.$$

For other values of q, we have to study the properties of the functions γ_q indirectly. In particular, these functions were studied by Polya [Po2].

If $0 < q \le 2$, then γ_q is a positive function which is, up to a constant 2π, equal to the density of the standard symmetric q-stable random variable; see this section below. We use Bernstein's theorem to prove the positivity of γ_q for these values of q.

A non-negative function f on $[0,\infty)$ is called *completely monotonic* if it is infinitely differentiable on $(0,\infty)$ and, for all $k \in \mathbb{N}$ and $x \in (0,\infty)$,

$$(-1)^k f^{(k)}(x) \ge 0.$$

The celebrated theorem of Bernstein (see [F, Ch. 18, Section 4]) states that every completely monotonic continuous at zero function is the Laplace transform of a finite measure on $[0,\infty)$, i.e. there exists a finite Borel measure μ on $[0,\infty)$ so that for every $x \ge 0$,

$$f(x) = \int_0^\infty e^{-tx}\,d\mu(t). \tag{2.47}$$

LEMMA 2.27. *For $0 < q \le 2$, the function γ_q is positive everywhere on $\mathbb{R}$.*

PROOF. It is well known and easy to compute that

$$(e^{-tz^2})^\wedge(\xi) = \sqrt{\pi}\,t^{-1/2}e^{-\xi^2/4t}. \tag{2.48}$$

This proves the lemma for $q = 2$. We also use this equality to prove the lemma for other values of q.

Let $0 < q < 2$. It is easily seen that, for every $s \in (0,1)$, the function e^{-z^s} is completely monotonic. Put $s = q/2 < 1$. Then by Bernstein's theorem, there exists a finite measure $\mu_{q/2}$ on $[0,\infty)$ so that for every $z \in [0,\infty)$

$$e^{-z^{q/2}} = \int_0^\infty e^{-tz}\,d\mu_{q/2}(t). \tag{2.49}$$

Hence, for every $z \in \mathbb{R}$,

$$e^{-|z|^q} = \int_0^\infty e^{-tz^2}\,d\mu_{q/2}(t). \tag{2.50}$$

Now we use (2.48) to compute the Fourier transform in both sides of the latter equation with respect to the variable z. Note that the function in the left-hand side of (2.50) is integrable, so one can use the Fubini theorem. We get that for every $\xi \in \mathbb{R}$,

$$\gamma_q(\xi) = \sqrt{\pi}\int_0^\infty t^{-1/2}e^{-\xi^2/4t}\,d\mu_{q/2}(t) > 0. \tag{2.51}$$

□

If q is not an even integer, the function $\gamma_q(t)$ decreases at infinity like $|t|^{-q-1}$; see [PS, Chapter 4, Problem 154].

LEMMA 2.28. *For any* $q > 0$

$$\lim_{t\to\infty} t^{1+q}\gamma_q(t) = 2\Gamma(q+1)\sin(\pi q/2).$$

PROOF. For $x > 0$, integrating by parts, we get

$$x^{q+1}\gamma_q(x) = 2x^{q+1}\int_0^\infty e^{-t^q}\cos(tx)\,dt = 2x^{q+1}\int_0^\infty \frac{\sin(tx)}{x}e^{-t^q}qt^{q-1}\,dt.$$

Making the change of variables $z = x^q t^q$ and putting $\delta = x^{-q}$ (note that $\delta \to 0$, as $x \to \infty$), we transform the latter integral into

$$2\,\Im\left(\int_0^\infty e^{-z\delta}e^{iz^{1/q}}\,dz\right).$$

Now use analyticity to change the line of integration from $[0,\infty)$ to $z = re^{i\theta}$ for some small $\theta < \pi q/2$. We get

$$2\,\Im\left(e^{i\theta}\int_0^\infty e^{-r\delta\cos\theta}e^{-ir\delta\sin\theta}e^{ir^{1/q}\cos(\theta/q)}e^{-r^{1/q}\sin(\theta/q)}\,dr\right).$$

Since $\sin(\theta/q) > 0$, we can use the dominated convergence theorem to pass to the limit, as $x \to \infty$ (or $\delta \to 0$.) We get

$$\lim_{x\to\infty} x^{1+q}\gamma_q(x) = 2\,\Im\left(e^{i\theta}\int_0^\infty e^{ir^{1/q}e^{i\theta/q}}\,dr\right).$$

To compute the latter integral, again use analyticity to change the line of integration from $z = re^{i\theta}$ to $z = re^{i\pi q/2}$. We get that the limit is equal to (use (2.26) and (2.27))

$$2\,\Im\left(e^{i\pi q/2}\int_0^\infty e^{-r^{1/q}}\,dr\right) = 2\Gamma(q+1)\sin(\pi q/2).$$

□

The result also holds if q is an even integer, when the limit is equal to zero. This means that, for even q, the function γ_q decreases even faster than t^{-1-q} (in fact, the rate is exponential).

We end this section by computing the moments of certain random variables, and, as a consequence, the moments of the function γ_q. If (Ω, P) is a probability space, then *random variables* on Ω are measurable real-valued functions on Ω. If f is a random variable, its *probability distribution* is a Borel measure μ on $\mathbb{R}$ such that for any Borel set $A \subset \mathbb{R}$,

$$\mu(A) = P\{\omega \in \Omega : \ f(\omega) \in A\}.$$

The *expectation* of a random variable is defined by

$$\mathbb{E}f = \int_\Omega f(\omega)\,dP(\omega) = \int_\mathbb{R} x\,d\mu(x).$$

The *characteristic functional* ϕ of a random variable is the Fourier transform of its probability distribution: for every $t \in R$,

$$\phi(t) = \mathbb{E}(e^{-itf}) = \int_\Omega e^{-itf(\omega)}dP(\omega) = \int_\mathbb{R} e^{-itx}d\mu(x).$$

We say that a random variable g is *normalized Gaussian* if for $t \in \mathbb{R}$, $\mathbb{E}(e^{-itg}) = e^{-\frac{t^2}{2}}$, a random variable η is *symmetric normalized q-stable* where $0 < q < 2$ if $\mathbb{E}(e^{-it\eta}) = e^{-|t|^q}$, and a positive random variable α is *normalized positive q-stable* for $0 < q < 1$ if $\mathbb{E}(e^{-t\alpha}) = e^{-t^q}$ when $t > 0$. The existence of symmetric q-stable random variables follows from Lemma 2.27, and the function $\gamma_q/(2\pi)$ is the density of the probability distribution of a symmetric normalized q-stable random variable. The existence of positive q-stable random variables follows from Bernstein's theorem, as in formula (2.49).

If g is normalized Gaussian, then the density of its probability distribution is $(1/\sqrt{2\pi})e^{-t^2/2}$, and it follows from (2.26) that

$$\mathbb{E}(|g|^z) = \frac{1}{\sqrt{\pi}} 2^{\frac{z}{2}} \Gamma(\frac{z+1}{2}), \qquad \Re z > -1. \tag{2.52}$$

The same argument as in Lemma 2.28 shows that the probability distribution of a positive normalized q-stable random variable α has density that decreases at infinity like t^{-1-q}, so the moments of this random variable exist up to the order q. To compute these moments, assume first that $-1 < \Re z < 0$. Then, by the definition of the Γ-function, we have

$$\mathbb{E}(\alpha^z) = \frac{1}{\Gamma(-z)} \int_0^\infty t^{-1-z} \mathbb{E}(e^{-t\alpha})\, dt$$

$$= \frac{1}{\Gamma(-z)} \int_0^\infty t^{-1-z} e^{-t^q} dt = \frac{\Gamma(-z/q)}{q\Gamma(-z)}. \tag{2.53}$$

Now we use analytic continuation to extend this formula to $-1 < \Re z < q$ (for $z = 0$, compute the limit).

Next, if η is normalized symmetric q-stable, then it has an identical distribution with $g\sqrt{2\alpha}$ where g, α are independent, g is normalized Gaussian and α is normalized positive $q/2$-stable. In fact, for every $t \in \mathbb{R}$,

$$\mathbb{E}(e^{-itg\sqrt{2\alpha}}) = \mathbb{E}(\exp(-t^2(\sqrt{2\alpha})^2/2)) = \mathbb{E}(e^{-t^2\alpha}) = e^{-|t|^q}. \tag{2.54}$$

By Lemma 2.28, the moments of η exist up to the order q. We can use (2.52) and (2.53) to compute these moments: by independence of g and α,

$$\mathbb{E}(|\eta|^z) = 2^{z/2}\mathbb{E}(g^z)\, \mathbb{E}(\alpha^{z/2}) = \frac{2^{z+1}\Gamma(\frac{-z}{q})\Gamma(\frac{z+1}{2})}{q\sqrt{\pi}\Gamma(-\frac{z}{2})}, \qquad -1 < \Re z < q. \tag{2.55}$$

In the special case of $q = 1$ formula (2.55) can be simplified either by using properties of the Γ-function or by a direct calculation, using Lemma 2.16 and the fact that the density of distribution of the symmetric 1-stable random variable ζ is $1/(\pi(1+x^2))$:

$$\mathbb{E}(|\zeta|^z) = \frac{1}{\pi} \int_{-\infty}^\infty \frac{|x|^z}{1+x^2} dx = (\cos(\frac{z\pi}{2}))^{-1}, \qquad -1 < \Re z < 1. \tag{2.56}$$

The latter formula allows us to rewrite (2.55) in a different form in the case $0 < q < 1$. If $0 < q < 1$ and η is normalized symmetric q-stable, then η has the same distribution as $\alpha\zeta$ where α, ζ are independent, α is normalized positive q-stable and ζ is normalized symmetric 1-stable (this argument is similar to (2.54)). Hence we can rewrite (2.55) in the form

$$\mathbb{E}(|\eta|^z) = \mathbb{E}(\alpha^z)\mathbb{E}(|\zeta|^z) = (\cos(\frac{z\pi}{2}))^{-1}\frac{\Gamma(\frac{-z}{q})}{q\Gamma(-z)}, \qquad -1 < \Re z < q. \tag{2.57}$$

Let us go back to our function γ_q. It follows from Lemma 2.28 that, for $-1 < z < q$, the moments

$$s_q(z) = \int_{\mathbb{R}} |t|^z \gamma_q(t)\, dt$$

converge absolutely. These moments can be computed precisely for all $q > 0$.

LEMMA 2.29. *For any $q > 0$ and $-1 < z < q$, where z is not an even integer, we have*

$$s_q(z) = \frac{2^{z+2}\sqrt{\pi}\Gamma(-z/q)\Gamma((z+1)/2)}{q\Gamma(-z/2)}.$$

In particular, if $q > 2$, the moments $s_q(z)$ are positive for $z \in (-1, 0)$ and $z \in (0, 2)$, and they are negative for $z \in (2, \min(q, 4))$.

PROOF. First assume that $0 < q \leq 2$ and $-1 < z < q$. Then the function γ_q is equal to 2π times the density of the normalized symmetric q-stable variable. Hence,

$$s_q(z) = 2\pi\ \mathbb{E}(|\eta|^z),$$

where η is a normalized symmetric q-stable random variable, and the result follows from (2.55).

Now suppose that $z \in (0, 2)$ is fixed and consider $s_q(z)$ as a function of q. This function of q is analytic on $z < \Re q$, which follows from direct differentiation by q. By analytic continuation, the formula for $s_q(z)$ holds for any $q > 2$ and $z \in (0, 2)$. Now fix $q > 2$ and note that $s_q(z)$ is an analytic function of z in $-1 < \Re z < q$. Again by analytic continuation the formula holds for all $q > 2$ and $-1 < z < q$, where z is not an even integer.

The sign of the number $s_q(z)$ can easily be determined from the signs of the corresponding values of the Γ-function; see Section 2.4. □

We now give an alternative proof of Lemma 2.29, which is an almost direct application of Parseval's formula.

PROOF. Assume that $-1 < z < 0$. We apply Parseval's formula, (2.34) with a non-integrable function $|t|^z$. To justify Parseval's formula in this case, one can use the definition of the Γ-function and (2.26) to write

$$t^z = \frac{2^{z+1/2}}{\Gamma(-z/2)} \int_0^\infty r^{-1-z} e^{-r^2t^2/2}\, dr,$$

which reduces the formula to the case of L_1-functions; see Lemma 3.21 for a similar argument.

By Lemma 2.23 with the coefficient appearing in (2.40), before simplifications:

$$\begin{aligned} s_q(z) &= \int_{\mathbb{R}} |t|^z \gamma_q(t)\, dt = \int_{\mathbb{R}} (|\cdot|^z)^\wedge(s) e^{-|s|^q}\, ds \\ &= \frac{2^{z+1}\sqrt{\pi}\Gamma((z+1)/2)}{\Gamma(-z/2)} \int_{\mathbb{R}} |s|^{-1-z} e^{-|s|^q}\, ds \\ &= \frac{2^{z+2}\sqrt{\pi}\Gamma(-z/q)\Gamma((z+1)/2)}{q\Gamma(-z/2)}. \end{aligned}$$

Now use that $s_q(z)$ is an analytic function of z in the domain $\{z \in \mathbb{C},\ -1 < \Re z < q\}$ to extend the formula to other values of z. $\square$

The result of Lemma 2.29 also shows that, for $q > 2$, the function γ_q is sign-changing, because its moments can be both positive and negative.

CHAPTER 3

Volume and the Fourier Transform

The Fourier analytic approach to sections of convex bodies is based on certain formulas expressing volume in terms of the Fourier transform. In this chapter we present several formulas of this kind.

3.1. The first examples: hyperplane sections of ℓ_q-balls

The results of this section will lead us to a general formula relating the volume of hyperplane sections of an arbitrary star body to the Fourier transform. As happened historically, we start with a formula for the volume of hyperplane sections of the cube. We denote by

$$B_\infty^n = \{x \in \mathbb{R}^n : \ \|x\|_\infty = \max_{k=1,\dots,n} |x_k| \le 1\}$$

the cube with side 2 in $\mathbb{R}^n$, which is the unit ball of the space ℓ_∞^n.

THEOREM 3.1. *For any* $\xi \in S^{n-1}$ *and* $t \in \mathbb{R}$,

$$A_{B_\infty^n,\xi}(t) = \mathrm{Vol}_{n-1}(B_\infty^n \cap \{\xi^\perp + t\xi\}) = \frac{2^n}{\pi}\int_0^\infty \cos(tr)\prod_{k=1}^n \frac{\sin(r\xi_k)}{r\xi_k}\,dr. \tag{3.1}$$

We assume here that $\sin(r\xi_k)/r\xi_k = 1$ *if* $\xi_k = 0$.

PROOF. The case where ξ has only one non-zero coordinate is trivial, so we assume that ξ has at least two non-zero coordinates. Then the function under the integral in (3.1) is integrable on $\mathbb{R}$.

Let χ be the indicator function of the interval $[-1,1]$. For every $x \in \mathbb{R}^n$,

$$\chi(\|x\|_\infty) = \prod_{k=1}^n \chi(|x_k|). \tag{3.2}$$

For any $r \in \mathbb{R}$, first using (2.1) and then the Fubini theorem and (3.2), we get

$$\widehat{A_\xi}(r) = \int_{\mathbb{R}} A_\xi(t)e^{-itr}\,dt = \int_{\mathbb{R}} e^{-itr}\left(\int_{(x,\xi)=t}\chi(\|x\|_\infty)\,dx\right)dt$$

$$= \int_{\mathbb{R}^n}\chi(\|x\|_\infty)e^{-ir(x,\xi)}\,dx = \prod_{k=1}^n\int_{-1}^1 e^{-irx_k\xi_k}\,dx_k = 2^n\prod_{k=1}^n\frac{\sin(r\xi_k)}{r\xi_k}.$$

Since the function A_ξ is even and $\widehat{A_\xi}$ is integrable on $\mathbb{R}$, we can invert the Fourier transform (see [Ru, Th 7.7]):

$$2\pi A_\xi(t) = (\widehat{A_\xi})^\wedge(t) = 2^n\int_{\mathbb{R}} e^{-itr}\prod_{k=1}^n\frac{\sin(r\xi_k)}{r\xi_k}\,dr.$$

□

A similar formula for ℓ_q-balls with $1 \le q \le 2$ was found by Meyer and Pajor, [MeP]. This formula was generalized to all $0 < q < \infty$ in [K7]. Let

$$B_q^n = \{x \in \mathbb{R}^n : \ \|x\|_q = \left(\sum_{k=1}^n |x_k|^q\right)^{1/q} \le 1\}$$

be the unit ball of the space ℓ_q^n, $0 < q < \infty$.

THEOREM 3.2. *For every $\xi \in S^{n-1}$ and every $q > 0$,*

$$\mathrm{Vol}_{n-1}(B_q^n \cap \xi^\perp) = \frac{q}{\pi(n-1)\Gamma((n-1)/q)} \int_0^\infty \prod_{k=1}^n \gamma_q(t\xi_k)\, dt,$$

where γ_q is the Fourier transform of the function $\exp(-|\cdot|^q)$ *on* $\mathbb{R}$.

We shall prove this formula in the next section, as a part of a more general result (use the formula of Theorem 3.8 and apply Lemma 3.6).

At this point, we would like to show that the formulas of Theorems 3.1 and 3.2 have something in common. To do this, let us compute the Fourier transform of the powers of ℓ_q^n-norms, $0 < q \le \infty$. We start with the case $q = \infty$, and we need the following simple fact.

LEMMA 3.3. *If $0 < p < n$, then the function*

$$g(\xi) = \int_0^\infty t^{-p-1} \prod_{k=1}^n |\frac{\sin(t\xi_k)}{\xi_k}|\, dt$$

is locally integrable on $\mathbb{R}^n$.

PROOF. Since the function g is homogeneous of degree $-n+p \in (-n, 0)$, it is enough to show that g is integrable on the unit cube Q_n in $\mathbb{R}^n$. We have

$$\int_{Q_n} g(\xi)\, d\xi = \int_0^\infty t^{-p-1} \left(\prod_{k=1}^n \int_{-1}^1 |\frac{\sin t\xi_k}{\xi_k}|\, d\xi_k\right) dt$$

$$= \int_0^\infty t^{-p-1} \left(\int_{-t}^t |\frac{\sin u}{u}|\, du\right)^n dt.$$

To see that the latter integral converges, break the outer integral into two integrals – over $[0, 1]$ and over $[1, \infty)$. For the first interval, we use the estimate

$$\int_{-t}^t |\frac{\sin u}{u}|\, du \le 2t,$$

and then we note that $n - p - 1 > -1$, so the integral converges.

For the second interval,

$$\int_{-t}^t |\frac{\sin u}{u}|\, du \le \int_{-1}^1 du + 2\int_1^t \frac{du}{u} = 2 + 2\ln t.$$

The integral over $[1, \infty)$ converges because $-1 - p < -1$. □

The function g is locally integrable and homogeneous of degree $-n+p$. By the argument of Lemma 2.1, for every test function ϕ on $\mathbb{R}^n$, the function $g(\xi)|\phi(\xi)|$ is integrable on $\mathbb{R}^n$, which justifies the use of Fubini's theorem in the next result. We now compute the Fourier transform of powers of the ℓ_∞-norm.

LEMMA 3.4. *If $p \in (0, n)$, then the Fourier transform of the function $\|\cdot\|_\infty^{-p}$ is equal to the locally integrable on $\mathbb{R}^n$ function*

$$\xi \mapsto 2^n p \int_0^\infty t^{-p-1} \prod_{k=1}^n \frac{\sin(t\xi_k)}{\xi_k}\, dt. \tag{3.3}$$

PROOF. For every $x \in \mathbb{R}^n$, $x \neq 0$, making the change of variables $u = z\|x\|_\infty$, we get

$$\|x\|_\infty^{-p} = p \int_0^\infty z^{p-1} \chi(z\|x\|_\infty)\, dz,$$

where χ is the indicator of $[-1, 1]$, and the integral converges because $p > 0$. Clearly,

$$\chi(z\|x\|_\infty) = \prod_{k=1}^n \chi(zx_k).$$

Therefore, for every fixed $z \neq 0$,

$$(\chi(z\|\cdot\|_\infty))^\wedge(\xi) = \prod_{k=1}^n \frac{2\sin(\xi_k/z)}{\xi_k}.$$

Since $0 < p < n$, for every test function $\phi \in \mathcal{S}(\mathbb{R}^n)$ the integral

$$\langle (\|\cdot\|_\infty^{-p})^\wedge, \phi\rangle = p \int_{\mathbb{R}^n} \hat{\phi}(x) \left(\int_0^\infty z^{p-1}\chi(z\|x\|_\infty)\, dz\right) dx \tag{3.4}$$

converges absolutely, and we can use the Fubini theorem, the definition of the Fourier transform of distributions and the change of variables $t = 1/z$ to show that the expression in the right-hand side of (3.4) is equal to

$$p \int_0^\infty z^{p-1} \langle \chi(z\|x\|_\infty), \hat{\phi}(x)\rangle\, dz$$

$$= p \int_0^\infty z^{p-1} \langle (\chi(z\|\cdot\|_\infty))^\wedge, \phi\rangle\, dz$$

$$= p \int_0^\infty z^{p-1} \left(\int_{\mathbb{R}^n} \prod_{k=1}^n \frac{2\sin(\xi_k/z)}{\xi_k} \phi(\xi)\, d\xi\right) dz$$

$$= 2^n p \int_{\mathbb{R}^n} \left(\int_0^\infty t^{-p-1} \prod_{k=1}^n \frac{\sin(t\xi_k)}{\xi_k}\, dt\right) \phi(\xi)\, d\xi$$

$$= 2^n p \Big\langle \int_0^\infty t^{-p-1} \prod_{k=1}^n \frac{\sin(t\xi_k)}{\xi_k}\, dt, \phi\Big\rangle.$$

□

In the next two lemmas we compute the Fourier transform of powers of ℓ_q-norms, $0 < q < \infty$. The necessary properties of the function γ_q can be found in Section 2.8.

LEMMA 3.5. *Let $q > 0$, $n \in \mathbb{N}$, $-n < p < 0$. Then the function*

$$h(\xi) = \int_0^\infty t^{n+p-1} \prod_{k=1}^n |\gamma_q(t\xi_k)|\, dt$$

is locally integrable on $\mathbb{R}^n$. If $0 < p < qn$, then the function h is integrable on compact sets outside of the coordinate planes in $\mathbb{R}^n$.

PROOF. First, let $-n < p < 0$. The function h is homogeneous of degree $-n-p$ so it is enough to prove that h is integrable on the unit cube Q_n in $\mathbb{R}^n$. We have

$$\int_{Q_n} h(\xi)\,d\xi = \int_0^\infty t^{p-1}\left(\int_{-t}^{t} |\gamma_q(z)|\,dz\right)^n dt.$$

When $t \in [0,1]$, we can estimate the function under the outer integral by

$$2^n t^{n+p-1}(\max_{t\in\mathbb{R}} |\gamma_q(t)|)^n.$$

Since $n+p-1 > -1$, the latter function is integrable on $[0,1]$. When $t \in [1,\infty)$, the function under the outer integral is bounded from above by $t^{p-1}\|\gamma_q\|^n_{L_1(\mathbb{R})}$, which is integrable on $[1,\infty)$ because $p < 0$. Note that $\gamma_q \in L_1(\mathbb{R})$ because it decreases at infinity as (or faster than) $|t|^{-1-q}$; see Lemma 2.28.

Let $0 < p < nq$. Since h is homogeneous of degree $-n-p$, it is enough to prove that h is integrable in the cube $[1,2]^n$. The integral of h over this cube is equal to

$$\int_0^\infty t^{p-1}\left(\int_t^{2t} |\gamma_q(z)|dz\right)^n dt.$$

For $t \in [0,1]$ the function under the outer integral can be estimated from above by $t^{p-1}\|\gamma_q\|^n_{L_1(\mathbb{R})}$, which is integrable on $[0,1]$, since now $p > 0$. By Lemma 2.28, there exists a constant C so that for every $t > 1$, $|\gamma_q(t)| \le C|t|^{-1-q}$ on $[t, 2t]$. This implies that the function under the outer integral is smaller than $C^n t^{p-1} t^{n(-1-q)} t^n = C^n t^{-1+p-qn}$, which is integrable on $[1,\infty)$ because $p < qn$. □

We now compute the Fourier transform of $\|\cdot\|_q^p$.

LEMMA 3.6. *Let $q > 0$, $n \in \mathbb{N}$, $-n < p < 0$. Then the Fourier transform of the distribution $\|\cdot\|_q^p$ is equal to a locally integrable function*

$$(\|\cdot\|_q^p)^\wedge(\xi) = \frac{q}{\Gamma(-p/q)} \int_0^\infty t^{n+p-1} \prod_{k=1}^n \gamma_q(t\xi_k)\,dt. \tag{3.5}$$

If $0 < p < nq$ and $p/q \notin \mathbb{N} \cup \{0\}$, the Fourier transform of $\|\cdot\|_q^p$ is equal to the same function outside of the coordinate planes in $\mathbb{R}^n$.

PROOF. First, suppose that $-n < p < 0$. Making the substitution $t = y\|x\|_q$ and using (2.26), we get

$$(|x_1|^q + \cdots + |x_n|^q)^{p/q} = \frac{q}{\Gamma(-p/q)} \int_0^\infty y^{-1-p} \exp(-y^q(|x_1|^q + \cdots + |x_n|^q))\,dy,$$

where the integral converges because $p < 0$.

For every fixed $y > 0$, the Fourier transform of the function

$$x \mapsto \exp(-y^q(|x_1|^q + \cdots + |x_n|^q))$$

at any point $\xi \in \mathbb{R}^n$ is equal to

$$y^{-n} \prod_{k=1}^n \gamma_q(\xi_k/y).$$

For any even test function ϕ, using Lemmas 3.5 and 2.1 to justify the application of Fubini's theorem and making the change of variables $t = 1/y$, we get

$$\langle(\|\cdot\|_q^p)^\wedge, \phi\rangle = \int_{\mathbb{R}^n} (|x_1|^q + \cdots + |x_n|^q)^{p/q} \hat{\phi}(x)\,dx$$

$$= \frac{q}{\Gamma(-p/q)} \int_0^\infty y^{-1-p} \Big\langle \exp(-y^q(|x_1|^q + \cdots + |x_n|^q)), \hat{\phi}(x) \Big\rangle dy$$

$$= \frac{q}{\Gamma(-p/q)} \int_0^\infty y^{-n-p-1} \left(\int_{\mathbb{R}^n} \prod_{k=1}^n \gamma_q(\xi_k/y)\phi(\xi)\, d\xi \right) dy$$

$$= \Big\langle \frac{q}{\Gamma(-p/q)} \int_0^\infty t^{n+p-1} \prod_{k=1}^n \gamma_q(t\xi_k)\, dt, \phi(\xi) \Big\rangle. \tag{3.6}$$

This proves the lemma for $-n < p < 0$.

By Lemma 2.22, the left-hand side of (3.6) is an analytic function of p in the domain $\Re p > -n$. Suppose that the test function ϕ in (3.6) is supported outside of the coordinate planes in $\mathbb{R}^n$. Then by Lemma 3.5 and an argument similar to that in Lemma 2.22 (recall that γ_q decreases at infinity like t^{-1-q}), the right-hand side of (3.6) is an analytic function of p in the domain $\Re p \in (-n, qn)$, $p/q \notin \mathbb{N} \cup \{0\}$. By the uniqueness of analytic continuation, equality (3.6) holds for all p from the latter domain. Since ϕ is an arbitrary test function supported outside of the coordinate planes, we get equality (3.5) outside of the coordinate planes in $\mathbb{R}^n$. □

Comparing the formulas of Theorems 3.1 and 3.2 with the results of Lemmas 3.4 and 3.6, one can notice that the volume of every central hyperplane section coincides (up to the same constant!) with the Fourier transform of $\|\cdot\|_q^{-n+1}$ at the corresponding point. This suggests that both formulas must be particular cases of a more general result.

3.2. A general formula for the volume of hyperplane sections

In this section we prove that the connection between the volume of sections of ℓ_q-balls and the Fourier transform, established in the previous section, can be extended to arbitrary symmetric star bodies.

LEMMA 3.7. *Let f be an even homogeneous function of degree $-n+1$ on $\mathbb{R}^n$, continuous on the sphere S^{n-1}. Then the Fourier transform of f is an even homogeneous of degree -1, continuous on $\mathbb{R}^n \setminus \{0\}$ function such that, for every $\xi \in S^{n-1}$,*

$$Rf(\xi) = \int_{S^{n-1}\cap\xi^\perp} f(\theta)d\theta = \frac{1}{\pi}\hat{f}(\xi),$$

where the spherical Radon transform R is applied to the restriction of f to the sphere.

PROOF. Since f is even, it is enough to consider even test functions ϕ. We have

$$\langle \hat{f}, \phi \rangle = \langle f, \hat{\phi} \rangle = \int_{\mathbb{R}^n} f(x)\hat{\phi}(x)\, dx$$

$$= \int_{S^{n-1}} f(\theta) \left(\int_0^\infty \hat{\phi}(t\theta)\, dt \right) d\theta. \tag{3.7}$$

Using Lemma 2.11 and the fact that the test function ϕ is even, we get that, for every fixed $\theta \in S^{n-1}$, the Fourier transform of the function $t \mapsto \hat{\phi}(t\theta)$ is equal to the function

$$z \mapsto 2\pi \int_{(x,\theta)=z} \phi(x)dx,$$

so

$$2\int_0^\infty \hat\phi(t\theta)\,dt = \int_{-\infty}^\infty \hat\phi(t\theta)dt$$
$$= (\hat\phi(t\theta))^\wedge(0) = 2\pi\int_{(x,\theta)=0}\phi(x)dx.$$

Substituting the latter integral in (3.7) and writing it in polar coordinates of the hyperplane $(x,\theta)=0$, we get that the expression in (3.7) is equal to

$$\pi\int_{S^{n-1}} f(\theta)\left(\int_{S^{n-1}\cap\theta^\perp}\left(\int_0^\infty r^{n-2}\phi(r\xi)dr\right)\,d\xi\right)\,d\theta$$

$$= \pi\int_{S^{n-1}} f(\theta)\;R\left(\int_0^\infty r^{n-2}\phi(r\xi)\,dr\right)(\theta)\,d\theta. \tag{3.8}$$

We now use self-duality of the spherical Radon transform (Lemma 2.13) with the functions $f(\theta)$ and $g(\xi)=\int_0^\infty r^{n-2}\phi(r\xi)dr$. We get that the expression in (3.8) is equal to

$$\pi\int_{S^{n-1}}\left(\int_0^\infty r^{n-2}\phi(r\xi)dr\right)Rf(\xi)\,d\xi$$
$$= \pi\int_{S^{n-1}}\left(\int_0^\infty r^{n-2}\phi(r\xi)dr\right)\left(\int_{S^{n-1}\cap\xi^\perp} f(\theta)d\theta\right)\,d\xi$$
$$= \pi\int_{\mathbb{R}^n}|x|_2^{-1}\left(\int_{S^{n-1}\cap(x/|x|_2)^\perp} f(\theta)d\theta\right)\phi(x)\,dx,$$

where, as before, $|\cdot|_2$ stands for the Euclidean norm in $\mathbb{R}^n$, and the last equality can be verified by passing to the polar coordinates $x=r\xi$.

We see now that the distribution $\hat f$ is equal to the locally integrable continuous on $\mathbb{R}^n\setminus\{0\}$ function

$$\pi|x|_2^{-1}\int_{S^{n-1}\cap(x/|x|_2)^\perp} f(\theta)d\theta.$$

In particular, for every unit vector $x\in S^{n-1}$, we get

$$\hat f(x) = \pi Rf(x).$$

□

An immediate consequence is a formula relating the volume of sections to the Fourier transform.

THEOREM 3.8. *Let K be an origin-symmetric star body in $\mathbb{R}^n$. The Fourier transform of the function $\|\cdot\|_K^{-n+1}$ is a homogeneous of degree -1 function on $\mathbb{R}^n$, continuous on $\mathbb{R}^n\setminus\{0\}$ and such that, for every $\xi\in S^{n-1}$,*

$$A_{K,\xi}(0) = \mathrm{Vol}_{n-1}(K\cap\xi^\perp) = \frac{1}{\pi(n-1)}(\|\cdot\|_K^{-n+1})^\wedge(\xi).$$

PROOF. Put $f=\|\cdot\|_K^{-n+1}$ in Lemma 3.7. By Lemma 3.7 and formula (2.17), $(\|\cdot\|_K^{-n+1})^\wedge$ is a continuous function outside of the origin in $\mathbb{R}^n$ and, for every $\xi\in S^{n-1}$,

$$(\|\cdot\|_K^{-n+1})^\wedge(\xi) = \pi R(\|\cdot\|_K^{-n+1})(\xi) = \pi(n-1)\mathrm{Vol}_{n-1}(K\cap\xi^\perp).$$

□

Theorem 3.8 provides a simple proof of Minkowski's uniqueness theorem for sections:

COROLLARY 3.9. *If K, L are origin-symmetric star bodies in $\mathbb{R}^n$ and for every direction $\xi \in S^{n-1}$,*

$$\mathrm{Vol}_{n-1}(K \cap \xi^\perp) = \mathrm{Vol}_{n-1}(L \cap \xi^\perp),$$

then $K = L$.

PROOF. By Theorem 3.8, we get that the Fourier transforms of homogeneous of degree $-n+1$ distributions $\|\cdot\|_K^{-n+1}$ and $\|\cdot\|_L^{-n+1}$ are equal on the whole $\mathbb{R}^n$. This implies that $\|x\|_K = \|x\|_L$ for every $x \in \mathbb{R}^n$, so $K = L$. □

We now prove a more general uniqueness theorem for the spherical Radon transform in arbitrary dimensions.

COROLLARY 3.10. *Let $1 \le m < n$ and let $f, g \in C(S^{n-1})$ be two even functions so that for any m-dimensional subspace H of $\mathbb{R}^n$,*

$$\int_{S^{n-1}\cap H} f(\theta)d\theta = \int_{S^{n-1}\cap H} g(\theta)d\theta. \tag{3.9}$$

Then $f = g$.

PROOF. Fix any $\xi \in S^{n-1}$ and denote by $G_\xi(n-1, m)$ the Grassmanian of m-dimensional subspaces of $\xi^\perp$. By (2.22), integrating both sides of equation (3.9) over $G_\xi(n-1, m)$, we get

$$\int_{S^{n-1}\cap\xi^\perp} f(\theta)d\theta = \int_{S^{n-1}\cap\xi^\perp} g(\theta)d\theta.$$

Extend f, g to homogeneous functions of degree $-n+1$ on the whole $\mathbb{R}^n$. By Lemma 3.7, the Fourier transforms of the extensions $f(\theta)r^{-n+1}$ and $g(\theta)r^{-n+1}$ are equal. By the uniqueness theorem for the Fourier transform, $f = g$. □

3.3. The parallel section function and the Fourier transform

The formula of Theorem 3.8 can be generalized in several different ways. The main result of this section, Theorem 3.18, shows that this formula is a part of a more general relation between the parallel section function and the Fourier transform. This relation serves as an important part of the solution of the Busemann-Petty problem.

We start with some properties of the Fourier transform of homogeneous functions. If f is an even continuous function on S^{n-1} and $p \in \mathbb{R}$, we denote by

$$f(\theta)r^p = f(x/|x|_2)|x|_2^p$$

an even homogeneous of degree p function of the variable $x = r\theta \in \mathbb{R}^n$, $r \in \mathbb{R}$, $\theta \in S^{n-1}$. When we write a function in the form $f(\theta)r^p$, we actually mean a function of the variable x, as in the formula above. If $p \neq -n-k$ for even non-negative integers k, then the action of the distribution $f(\theta)r^p$ on a test function ϕ can be written as

$$\begin{aligned}\langle f(\theta)r^p, \phi\rangle &= \langle f(x/|x|_2)|x|_2^p, \phi(x)\rangle \\ &= \frac{1}{2}\int_{S^{n-1}} f(\theta)\langle |z|^{n+p-1}, \phi(z\theta)\rangle d\theta.\end{aligned}$$

Moreover, if $p > -n$ or if ϕ is supported outside of the origin, then the right-hand side of the latter equation can be written as

$$\frac{1}{2}\int_{S^{n-1}} f(\theta)\left(\int_{\mathbb{R}} |z|^{n+p-1}\phi(z\theta)dz\right)d\theta.$$

We know from Lemma 2.21 that the Fourier transform of a homogeneous function of degree p is a homogeneous distribution of degree $-n-p$. We want to know when this Fourier transform is a continuous function on $\mathbb{R}^n \setminus \{0\}$. We also want to study the convergence of such Fourier transforms.

LEMMA 3.11. *(i) Suppose that $p > -n$ and let f_k, $k \in \mathbb{N}$, and f be even continuous functions on the sphere S^{n-1} so that $f_k \to f$ in $C(S^{n-1})$. Then for every even test function ϕ,*

$$\lim_{k\to\infty}\langle (f_k(\theta)r^p)^\wedge, \phi\rangle = \langle (f(\theta)r^p)^\wedge, \phi\rangle.$$

(ii) Suppose that $\delta > 0$, a sequence of numbers $\{p_k\}_{k=1}^\infty$ is such that $p_k > -n+\delta$ for all k, and $\lim_{k\to\infty} p_k = p$. Let $f \in C(S^{n-1})$. Then for every even test function ϕ,

$$\lim_{k\to\infty}\langle (f(\theta)r^{p_k})^\wedge, \phi\rangle = \langle (f(\theta)r^p)^\wedge, \phi\rangle.$$

PROOF. The functions f_k are uniformly bounded by some constant C. Since $p > -n$, by the same argument as in Lemma 2.1, the functions $f_k(\theta)r^p$ are locally integrable on $\mathbb{R}^n$, and the functions $f_k(\theta)r^p\hat{\phi}(x)$ are integrable on $\mathbb{R}^n$. These functions are majorated by an integrable function $Cr^p|\hat{\phi}(x)|$ and converge pointwise to $f(\theta)r^p\hat{\phi}(x)$. By the Lebesgue dominated convergence theorem,

$$\lim_{k\to\infty}\langle (f_k(\theta)r^p)^\wedge, \phi\rangle = \lim_{k\to\infty}\int_{\mathbb{R}^n} f_k(\theta)r^p\hat{\phi}(x)dx$$

$$= \int_{\mathbb{R}^n} f(\theta)r^p\hat{\phi}(x)dx = \langle (f(\theta)r^p)^\wedge, \phi\rangle,$$

and we are done with the first part.

For the second part, the functions $f(\theta)r^{p_k}$ are locally integrable, since each $p_k > -n$. Hence, the functions $f(\theta)r^{p_k}\hat{\phi}(x)$ are integrable on $\mathbb{R}^n$. There exists a constant $s > -n$ so that $p_k < s$ for each k. Now the functions r^{p_k} are majorated by the function $h(r) = \max(r^s, r^{-n+\delta})$. Therefore, the functions $f(\theta)r^{p_k}\hat{\phi}(x)$ are majorated by an integrable function $f(\theta)h(r)|\hat{\phi}(x)|$, and they converge pointwise to $f(\theta)r^p\hat{\phi}(x)$. We can finish the proof in the same way as in the first part. □

The following formula is well known:

LEMMA 3.12. *For every $q > -1$ and every $x \in S^{n-1}$, we have*

$$\int_{S^{n-1}} |(x,\xi)|^q\, d\xi = \frac{2\Gamma((q+1)/2)\pi^{(n-1)/2}}{\Gamma((n+q)/2)}.$$

PROOF. It is easily seen that for any integrable function Φ on $[-1,1]$ and any $x \in S^{n-1}$,

$$\int_{S^{n-1}} \Phi((x,\xi))\, d\xi = |S^{n-2}|\int_{-1}^{1} \Phi(t)(1-t^2)^{(n-3)/2}\, dt,$$

where $|S^{n-2}|$ is the surface area of the unit Euclidean ball in $\mathbb{R}^{n-1}$ (see for example [Gr, p. 9]), and the measure $d\xi$ on the sphere has density 1 and is not normalized.

Putting $\Phi(t) = |t|^q$ (note that $q > -1$ so the integral converges) and using formula (2.31), we get the result. □

LEMMA 3.13. *If $q > -1$ and b is a continuous function on the sphere S^{n-1}, then the function*

$$h(x) = \int_{S^{n-1}} |(x,\xi)|^q b(\xi)\, d\xi$$

is continuous on S^{n-1}. Also,

$$\|h\|_{C(S^{n-1})} \le \frac{2\Gamma((q+1)/2)\pi^{(n-1)/2}}{\Gamma((n+q)/2)} \|b\|_{C(S^{n-1})}.$$

PROOF. Let x_k be a sequence of points in the sphere converging to $x \in S^{n-1}$. Then one can find a sequence of rotations T_k such that $T_k(x) = x_k$, and

$$\|b(T_k) - b\|_{C(S^{n-1})} \to 0,$$

as $k \to \infty$ (b is uniformly continuous on the compact set S^{n-1}). Then,

$$h(x_k) - h(x) = \int_{S^{n-1}} |(x,\xi)|^q (b(T_k\xi) - b(\xi))\, d\xi,$$

and the result follows from Lemma 3.12. □

Our next lemma is a combination of Lemmas 2.23 and 2.11. The general idea behind this lemma is quite simple. If $f \in L_1(\mathbb{R})$ and $\xi \in S^{n-1}$, consider a measure μ in $\mathbb{R}^n$ supported in the line $t\xi$, $t \in \mathbb{R}$, with density f. The Fourier transform of the measure μ is the function

$$x \mapsto \hat{f}((x,\xi)),\ x \in \mathbb{R}^n.$$

In fact,

$$\hat{\mu}(x) = \int_{\mathbb{R}^n} e^{-i(x,\eta)}\, d\mu(\eta) = \int_{\mathbb{R}} f(t) e^{-it(x,\xi)}\, dt = \hat{f}((x,\xi)). \tag{3.10}$$

Therefore for any test function ϕ,

$$\int_{\mathbb{R}^n} \hat{f}((x,\xi))\phi(x)\, dx = \langle \hat{\mu}, \phi\rangle = \langle \mu, \hat{\phi}\rangle = \int_{\mathbb{R}} f(t)\hat{\phi}(t\xi)\, dt.$$

We would like to have a similar formula in the case where f is not necessarily an L_1-function.

LEMMA 3.14. *Let $q > -1$, where q is not an even integer. Then for every even test function $\phi \in \mathcal{S}(\mathbb{R}^n)$ supported outside of the origin and every fixed vector $\xi \in \mathbb{R}^n$, $\xi \ne 0$, we have*

$$\int_{\mathbb{R}^n} |(x,\xi)|^q \hat{\phi}(x)\, dx = -2(2\pi)^{n-1}\Gamma(q+1)\sin(\pi q/2) \int_{\mathbb{R}} |t|^{-1-q}\phi(t\xi)\, dt.$$

PROOF. Because of homogeneity, it is enough to prove the lemma for $\xi \in S^{n-1}$. Since ϕ is an even function, we have $(\hat{\phi})^\wedge = (2\pi)^n\phi$. By Lemma 2.11, the function $t \mapsto (2\pi)^n\phi(t\xi)$ is the Fourier transform of the function $z \mapsto \int_{(x,\xi)=z} \hat{\phi}(x)\, dx$. Using the Fubini theorem and Lemma 2.23, we get

$$\int_{\mathbb{R}^n} |(x,\xi)|^q\, \hat{\phi}(x)\, dx = \int_{\mathbb{R}} |z|^q \left(\int_{(x,\xi)=z} \hat{\phi}(x)\, dx \right) dz$$

$$= \left\langle |z|^q, \int_{(x,\xi)=z} \hat{\phi}(x)\,dx \right\rangle$$

$$= \frac{1}{2\pi}\left\langle (-2\Gamma(1+q)\sin(\pi q/2))|t|^{-1-q}, (2\pi)^n\phi(t\xi) \right\rangle$$

$$= (2\pi)^{n-1}(-2\Gamma(1+q)\sin(\pi q/2)) \int_{\mathbb{R}} |t|^{-1-q}\phi(t\xi)\,dt.$$

□

COROLLARY 3.15. *Suppose that $q > -1$ is not an even integer and b is an even continuous function on the sphere S^{n-1}. Define a function F on $\mathbb{R}^n$ by*

$$F(x) = \int_{S^{n-1}} |(x,\xi)|^q\, b(\xi)\, d\xi.$$

Then the Fourier transform $\widehat{F}$ is a homogeneous of degree $-n-q$ continuous on $\mathbb{R}^n \setminus \{0\}$ function such that, for every $y \in \mathbb{R}^n \setminus \{0\}$,

$$\widehat{F}(y) = -4(2\pi)^{n-1}\Gamma(q+1)\sin(\pi q/2)\, b(\xi) r^{-n-q},$$

where $r = |y|_2$ and $\xi = y/|y|_2$.

PROOF. For any even test function ϕ supported outside of the origin, using Lemma 3.14, we get

$$\langle \widehat{F}, \phi\rangle = \langle F, \hat{\phi}\rangle = \int_{S^{n-1}} b(\xi) \left(\int_{\mathbb{R}^n} |(x,\xi)|^q \hat{\phi}(x)\, dx \right) d\xi$$

$$= -2(2\pi)^{n-1}\Gamma(q+1)\sin(\pi q/2) \int_{S^{n-1}} b(\xi) \left(\int_{\mathbb{R}} |t|^{-1-q}\phi(t\xi)\, dt \right) d\xi$$

$$= -4(2\pi)^{n-1}\Gamma(q+1)\sin(\pi q/2)\, \langle b(\xi) r^{-n-q}, \phi\rangle.$$

□

The next lemma shows that the Fourier transform of a smooth homogeneous function is also smooth. It is possible to prove a stronger result with exact degrees of smoothness of a function and its Fourier transform, but this would make the text more complicated, and, on the other hand, we do not really need it here.

LEMMA 3.16. *Let $k \in \mathbb{N} \cup \{0\}$ and let $f \in C^{2k}(S^{n-1})$ be an even function. Suppose that $q \le 2k$, where q is not an odd integer. Then the following hold.*
(i) The Fourier transform of the distribution $f(\theta)r^{-n+q+1}$ is a homogeneous of degree $-1-q$, continuous on $\mathbb{R}^n \setminus \{0\}$ function. If $q < 2k$, then for every $x \in \mathbb{R}^n$,

$$|x|_2^{2k}(f(\theta)r^{-n+q+1})^\wedge(x)$$

(3.11)
$$= \frac{(-1)^k\pi}{-2\Gamma(2k-q)\sin(\pi(2k-q-1)/2)} \int_{S^{n-1}} |(x,\xi)|^{2k-q-1}\Delta^k(f(\theta)r^{-n+q+1})(\xi)\, d\xi.$$

If $q = 2k$, then

$$|x|_2^{2k}\ (f(\theta)r^{-n+q+1})^\wedge(x)$$

$$= (-1)^k\pi|x|_2^{-1} \int_{S^{n-1}\cap(x/|x|_2)^\perp} \Delta^k(f(\theta)r^{-n+q+1})(\xi)\, d\xi. \tag{3.12}$$

Here $\Delta = \sum_{i=1}^n \partial^2/\partial x_i^2$ is the Laplace operator on $\mathbb{R}^n$.

(ii) If $f \in C^\infty(S^{n-1})$, then there exists an even function $g \in C^\infty(S^{n-1})$ so that for every $x = t\xi \in \mathbb{R}^n$, $t \neq 0$, $\xi \in S^{n-1}$,

$$(f(\theta)r^{-n+q+1})^\wedge(x) = g(\xi)t^{-1-q},$$

so the Fourier transform of $f(\theta)r^{-n+q+1}$ is an infinitely smooth function on $\mathbb{R}^n\setminus\{0\}$.

PROOF. For any even test function ϕ such that $\hat{\phi}$ is supported outside of the origin, by the connection between differentiation and the Fourier transform, formula (2.36),

$$\langle |x|_2^{2k}(f(\theta)r^{-n+q+1})^\wedge(x), \phi(x)\rangle = (-1)^k\langle \Delta^k(f(\theta)r^{-n+q+1}), \hat{\phi}\rangle.$$

On the other hand, if $q < 2k$, where q is not an odd integer, then, by Corollary 3.15,

$$\Big\langle \int_{S^{n-1}} |(x,\xi)|^{2k-q-1}\Delta^k(f(\theta)r^{-n+q+1})(\xi)\, d\xi, \phi(x)\Big\rangle$$

$$= \frac{1}{(2\pi)^n}\Big\langle \Big(\int_{S^{n-1}} |(x,\xi)|^{2k-q-1}\Delta^k(f(\theta)r^{-n+q+1})(\xi)\, d\xi\Big)^\wedge (u), \hat{\phi}(u)\Big\rangle$$

$$= \frac{-4(2\pi)^{n-1}\Gamma(2k-q)\sin(\pi(2k-q-1)/2)}{(2\pi)^n}\langle \Delta^k(f(\theta)r^{-n+q+1}), \hat{\phi}\rangle,$$

because $\Delta^k(f(\theta)r^{-n+q+1})$ is a homogeneous function of degree $-n-2k+q+1$, which is continuous on $\mathbb{R}^n \setminus \{0\}$. This shows that the functions in both sides of (3.11), considered as distributions, coincide on test functions ϕ whose Fourier transform is supported outside of the origin. Such distributions can differ only by a polynomial. But this polynomial must be zero because both functions are even homogeneous of degree $2k - q - 1$, which is not an even integer.

If $q = 2k$, the function $\Delta^k(f(\theta)r^{-n+q+1})$ is homogeneous of degree $-n + 1$. Formula (3.12) follows from Lemma 3.7.

Finally, if f is infinitely smooth, then we can choose k in (3.11) as large as we want. The function of x in the right-hand side of (3.11) is $[2k - q - 3]$ times differentiable by the variable x. To see this, first note that for $p > 1$

$$\frac{\partial}{\partial x_i}|(x,\xi)|^p = p\xi_i|(x,\xi)|^{p-1}\mathrm{sgn}((x,\xi)),$$

$$\frac{\partial}{\partial x_i}(|(x,\xi)|^p\mathrm{sgn}((x,\xi))) = p\xi_i|(x,\xi)|^{p-1},$$

and then use a standard justification of differentiation under the integral (the best way to do it is backwards, integrating the derivative and using the Fubini theorem; see [K19] for similar calculations). □

An immediate consequence of Lemma 3.16 and Lemma 3.13 is the following.

COROLLARY 3.17. *Suppose that $q \leq 2k$, $k \in \mathbb{N} \cup \{0\}$.*
(i) If $f \in C^{2k}(S^{n-1})$, then the Fourier transform $(f(\theta)r^{-n+q+1})^\wedge$ is a homogeneous of degree $-1 - q$ continuous on $\mathbb{R}^n \setminus \{0\}$ function.
(ii) If $f_m, f \in C^{2k}(S^{n-1})$ and f_m converges to f in $C^{2k}(S^{n-1})$, then the Fourier transforms $(f_m(\theta)r^{-n+q+1})^\wedge$ converge to $(f(\theta)r^{-n+q+1})^\wedge$ in the space $C(S^{n-1})$.

Let D be an infinitely smooth origin-symmetric convex body in $\mathbb{R}^n$. By Lemma 2.4, for any $m \in \mathbb{N} \cup \{0\}$, there exists an interval $U_m = (-\delta_m, \delta_m) \subset \mathbb{R}$ so that the parallel section functions $A_{D,\xi}$ are uniformly differentiable up to the order m in this interval, and, for any $t \in U_m$ and $0 \le k \le m$, the function $\xi \mapsto A_{D,\xi}^{(k)}(t)$ is continuous on S^{n-1}.

Applying the definition of fractional derivatives (2.41) to the functions $A_{D,\xi}$ and changing the interval $(0,1)$ to $(0,\delta_m)$, we get that for $\Re q < m$, $q \ne 0, 1, ..., m-1$,

$$A_\xi^{(q)}(0) = \frac{1}{\Gamma(-q)} \int_0^{\delta_m} t^{-1-q} \left(A_\xi(t) - A_\xi(0) - ... - A_\xi^{(m-1)}(0) \frac{t^{m-1}}{(m-1)!} \right) dt$$

$$+ \frac{1}{\Gamma(-q)} \int_{\delta_m}^{\infty} t^{-1-q} A_\xi(t)\, dt + \frac{1}{\Gamma(-q)} \sum_{k=0}^{m-1} \frac{\delta_m^{k-q} A_\xi^{(k)}(0)}{k!(k-q)}.$$

Using a formula for the remainder of the Taylor series and an argument similar to that from Lemma 2.22, we see that the fractional derivative $A_{D,\xi}^{(q)}(0)$ is an analytic function of q everywhere in $\mathbb{C}$ (we have to extend the definition to integers, as usual). Moreover, $A_{D,\xi}^{(q)}(0)$ is a continuous function of ξ on the sphere. Extending this function to a homogeneous function of ξ of degree $-1-q$, one can see by the argument of Lemma 2.22 that, for any test function ϕ on $\mathbb{R}^n$, the function

$$q \mapsto \langle A_{D,\xi}^{(q)}(0), \phi(\xi) \rangle$$

is analytic in the domain $\{q \in \mathbb{C} : -1 < \Re q\}$.

We are now ready to prove the main result of this section expressing the derivatives of the parallel section function $A_{D,\xi}$ in terms of the Fourier transform of powers of the Minkowski functional. In the next theorem, by Corollary 3.17, the Fourier transform of $\|\cdot\|_D^{-n+q+1}$ is a homogeneous of degree $-1-q$ continuous function on $\mathbb{R}^n \setminus \{0\}$, which allows us to write all the formulas pointwise on the sphere.

THEOREM 3.18. *Let D be an infinitely smooth origin-symmetric convex body in $\mathbb{R}^n$, $\xi \in S^{n-1}$. Then, for every $q \in (-1, \infty)$, $q \ne n-1$, the fractional derivative of the order q of the parallel section function at zero can be expressed in the form*

$$A_{D,\xi}^{(q)}(0) = \frac{\cos(\pi q/2)}{\pi(n-q-1)} (\|\cdot\|_D^{-n+q+1})^\wedge(\xi). \tag{3.13}$$

In particular, if $k \ge 0$, $k \ne n-1$, is an even integer, then we get the usual derivative of the order k :

$$(\|\cdot\|_D^{-n+k+1})^\wedge(\xi) = (-1)^{k/2} \pi (n-k-1) A_{D,\xi}^{(k)}(0). \tag{3.14}$$

If $k \ge 1$, $k \ne n-1$, is an odd integer, then

$$(\|\cdot\|_D^{-n+k+1})^\wedge(\xi) = (-1)^{(k+1)/2} 2(n-k-1)k!$$

$$\times \int_0^\infty \frac{A_{D,\xi}(z) - A_{D,\xi}(0) - ... - A_{D,\xi}^{(k-1)}(0) \frac{z^{k-1}}{(k-1)!}}{z^{k+1}}\, dz. \tag{3.15}$$

PROOF. Let us start with $q \in (-1, 0)$. Using (2.41), (2.1) and the Fubini theorem, we get (as before, χ is the indicator of $[-1,1]$)

$$A_{D,\xi}^{(q)}(0) = \frac{1}{2\Gamma(-q)} \int_{-\infty}^{\infty} |z|^{-q-1} A_{D,\xi}(z)\, dz$$

$$= \frac{1}{2\Gamma(-q)} \int_{-\infty}^{\infty} |z|^{-q-1} \left(\int_{(x,\xi)=z} \chi(\|x\|_D)\, dx \right) dz$$

$$= \frac{1}{2\Gamma(-q)} \int_{\mathbb{R}^n} |(x,\xi)|^{-q-1} \chi(\|x\|_D)\, dx$$

$$= \frac{1}{2\Gamma(-q)} \int_{S^{n-1}} |(\theta,\xi)|^{-q-1} \left(\int_0^{\infty} r^{n-q-2} \chi(r\|\theta\|_D)\, dr \right) d\theta$$

$$= \frac{1}{2(n-q-1)\Gamma(-q)} \int_{S^{n-1}} |(\theta,\xi)|^{-q-1} \|\theta\|_D^{-n+q+1}\, d\theta.$$

We now consider $A_{D,\xi}^{(q)}(0)$ as a function of $\xi \in \mathbb{R}^n \setminus \{0\}$ so that it is homogeneous of degree $-1-q$. For every even test function ϕ, using Lemma 3.14, we get

$$\langle A_{D,\xi}^{(q)}(0), \phi(\xi) \rangle = \frac{1}{2\Gamma(-q)(n-q-1)}$$

$$\times \int_{S^{n-1}} \|\theta\|_D^{-n+q+1} \left(\int_{\mathbb{R}^n} |(\theta,\xi)|^{-q-1} \phi(\xi)\, d\xi \right) d\theta$$

$$= \frac{-1}{4(n-q-1)\Gamma(-q)\Gamma(q+1)\sin\frac{q\pi}{2}}$$

$$\times \int_{S^{n-1}} \|\theta\|_D^{-n+q+1} \left(\int_{\mathbb{R}} |t|^q\, \widehat{\phi}(t\theta)\, dt \right) d\theta$$

$$= \frac{\cos(q\pi/2)}{\pi(n-q-1)} \left\langle (\|\cdot\|_D^{-n+q+1})^\wedge(\xi), \phi(\xi) \right\rangle, \tag{3.16}$$

where the last equation follows from (2.32) and the simple calculation

$$\langle \left(\|\cdot\|_D^{-n+q+1} \right)^\wedge, \phi \rangle = \int_{\mathbb{R}^n} \|x\|_D^{-n+q+1} \hat{\phi}(x)\, dx$$

$$= \int_{S^{n-1}} \|\theta\|_D^{-n+q+1} \left(\int_0^{\infty} r^q\, \hat{\phi}(r\theta)\, dr \right) d\theta$$

$$= \frac{1}{2} \int_{S^{n-1}} \|\theta\|_D^{-n+q+1} \left(\int_{\mathbb{R}} |r|^q\, \hat{\phi}(r\theta)\, dr \right) d\theta.$$

By the remark before Theorem 3.18 and by Lemma 2.22, both sides of (3.16) are analytic functions of q in the domain $-1 < \Re q$, $q \neq n-1$. By analytic extension, this proves (3.16) for all required values of q. Also, by the same remark before the theorem, $A_{D,\xi}^{(q)}(0)$ is a continuous function of ξ on the sphere, and by Corollary 3.17 so is the Fourier transform of $\|\cdot\|_D^{-n+q+1}$, so we get equality (3.13) as the equality of functions on the sphere. If k is even, we immediately get (3.14).

However, if k is odd, both sides of (3.13) vanish (recall that D is origin-symmetric, so $A_{D,\xi}$ is an even function). Since we have already proved (3.16) for $k-1 < q < k$, we divide both sides of (3.16) by $\cos(\pi q/2)$ and compute the limit as $q \to k$. Here we use expression (2.42) for the fractional derivative. The limit in the left-hand side of (3.16) is equal to the action on the test function $\phi(\xi)$ of the following function of ξ :

$$\lim_{q\to k} \frac{1}{\Gamma(-q)\cos(\pi q/2)}$$

$$\times \int_0^\infty \frac{A_{D,\xi}(z) - A_{D,\xi}(0) - A''_{D,\xi}(0)\frac{z^2}{2} - \cdots - A_{D,\xi}^{(k-1)}(0)\frac{z^{k-1}}{(k-1)!}}{z^{k+1}}\, dz.$$

Now,

$$\begin{aligned}
&\lim_{q\to k} \Gamma(-q)\cos(\pi q/2)\\
&= \lim_{q\to k} \Gamma(-q) \sin\left(\frac{(q+1)\pi}{2} - \frac{(k+1)\pi}{2}\right)(-1)^{(k+1)/2}\\
&= \lim_{q\to k} \frac{\Gamma(-q+k+1)}{(-q)(1-q)\cdots(k-q)} \sin((q-k)\pi/2)(-1)^{(k+1)/2}\\
&= -\frac{\pi}{2}(-1)^{(k+1)/2}(-1)^k\frac{1}{k!} = \frac{\pi}{2}(-1)^{(k+1)/2}\frac{1}{k!}.
\end{aligned}$$

On the other hand, by Lemma 3.11,

$$\lim_{q\to k} \langle (\|\cdot\|_D^{-n+q+1})^\wedge(\xi), \phi(\xi)\rangle = \langle (\|\cdot\|_D^{-n+k+1})^\wedge(\xi), \phi(\xi)\rangle,$$

and the result in the case of odd integers follows. □

3.4. Parseval's formula on the sphere

In this section we prove a spherical version of Parseval's formula. Let us start with an informal reasoning that gives an idea of how this can be done by projecting the classical formula to the sphere.

Suppose that K and L are star bodies in $\mathbb{R}^n$, $0 < p < n$, and f is an even function on $\mathbb{R}$. Consider an integral

$$\int_{\mathbb{R}^n} \|x\|_K^{-p} f(\|x\|_L)\, dx. \tag{3.17}$$

Writing this integral in polar coordinates and making the change of variables $t = r\|\theta\|_L$, we see that (3.17) is equal to

$$\int_{S^{n-1}} \|\theta\|_K^{-p}\left(\int_0^\infty r^{n-p-1} f(r\|\theta\|_L)\, dr\right) d\theta$$

$$= \left(\int_0^\infty t^{n-p-1} f(t)\, dt\right)\left(\int_{S^{n-1}} \|\theta\|_K^{-p}\|\theta\|_L^{-n+p}\, d\theta\right). \tag{3.18}$$

On the other hand, applying the classical Parseval formula (informally), we get that the integral (3.17) is also equal to

$$\frac{1}{(2\pi)^n}\int_{\mathbb{R}^n} (\|\cdot\|_K^{-p})^\wedge(\xi)(f(\|\cdot\|_L))^\wedge(\xi)\, d\xi.$$

Now recall that the Fourier transform of a homogeneous function of degree $-p$ is a homogeneous function of degree $-n+p$ and write the latter integral in polar coordinates $\xi = r\theta$. We get that (3.17) equals

$$\frac{1}{(2\pi)^n}\int_{S^{n-1}} (\|\cdot\|_K^{-p})^\wedge(\theta)\left(\int_0^\infty r^{p-1}(f(\|\cdot\|_L))^\wedge(r\theta)\, dr\right) d\theta. \tag{3.19}$$

The crucial observation is that

$$\int_0^\infty r^{p-1}(f(\|\cdot\|_L))^\wedge(r\theta)\, dr = c_p(f)(\|\cdot\|_L^{-n+p})^\wedge(\theta), \tag{3.20}$$

where $c_p(f)$ is a constant not depending on θ. This means that directional moments of the Fourier transform of functions of the form $f(\|\cdot\|_L)$ do not depend on the choice of f (up to a constant independent of the direction). Combining (3.20) with (3.19) and (3.18), we get a spherical version of Parseval's formula, which is the main goal of this section:

$$\int_{S^{n-1}} (\|\cdot\|_K^{-p})^\wedge(\theta)(\|\cdot\|_L^{-n+p})^\wedge(\theta)\, d\theta = d_p(f)\int_{S^{n-1}} \|\theta\|_K^{-p}\|\theta\|_L^{-n+p}\, d\theta. \tag{3.21}$$

To make this reasoning formally correct, we have to do several things. First, the Fourier transforms of $\|\cdot\|_K^p$ and $\|\cdot\|_L^{-n+p}$ must be continuous functions on the sphere. This can be achieved by assuming that the bodies K and L are sufficiently smooth. Secondly, the integrals involving the function f must converge. To ensure this, we need to choose a specific function f so that the derivatives of the function $f(\|\cdot\|_L)$ of high enough orders are integrable on $\mathbb{R}^n$. Then, using the connection between differentiation and the Fourier transform, we get that the Fourier transform of $f(\|\cdot\|_L)$ is quickly decreasing at infinity. For example, if

$$\Delta^{(n+2)/2}(f(\|\cdot\|_L)) \in L_1(\mathbb{R}^n), \tag{3.22}$$

then, by (2.36),

$$|(f(\|\cdot\|_L))^\wedge(x)| \le M|x|_2^{-n-2},$$

where M is the L_1-norm of the function (3.22). We are also going to show that, as one can expect, the constant $d_p(f) = (2\pi)^n$, and, thus, it does not depend on f or p.

Before we proceed with choosing a specific even function f, let us also informally deduce (3.20). Writing the Fourier integral in polar coordinates $\xi = r\eta$ and making a change of variables $t = r\|\eta\|_L$, we get

$$(f(\|\cdot\|_L))^\wedge(\xi) = \int_{\mathbb{R}^n} f(\|x\|_L)e^{-i(x,\xi)}\, dx$$

$$= \int_{S^{n-1}} \left(\int_0^\infty r^{n-1} f(r\|\eta\|_L)e^{-ir(\eta,\xi)}dr\right) d\eta$$

$$= \frac{1}{2}\int_{S^{n-1}} \|\eta\|_L^{-n}(|t|^{n-1}f(t))^\wedge((\eta,\xi)/\|\eta\|_L)\, d\eta.$$

Putting $\xi = r\theta$ and integrating with respect to $r^{p-1}dr$, we see that the left-hand side of (3.20) equals

$$\left(\int_0^\infty r^{p-1}(|t|^{n-1}f(t))^\wedge(r)\, dr\right)\left(\int_{S^{n-1}} \|\eta\|_L^{-n+p}|(\eta,\theta)|^{-p}\, d\eta\right).$$

On the other hand (and also informally),

$$(\|\cdot\|_L^{-n+p})^\wedge(\theta) = \int_{\mathbb{R}^n} \|x\|_L^{-n+p}e^{-i(x,\theta)}dx$$

$$= \int_{S^{n-1}} \|\eta\|_L^{-n+p}\left(\int_0^\infty r^{p-1}e^{-ir(\eta,\theta)}dr\right) d\eta$$

$$= c_{p-1}\int_{S^{n-1}} \|\eta\|_L^{-n+p}|(\eta,\theta)|^{-p}\, d\eta,$$

which expresses the right-hand side of (3.20) (up to a constant) in exactly the same way as the left-hand side. Of course the integrals in the latter argument do not converge, which will be corrected below by using test functions.

Let us make this reasoning work. We put $f(t) = e^{-t^4}$. For an origin-symmetric star body K in $\mathbb{R}^n$, define a function

$$\mu_K(\xi) = (\exp(-\|\cdot\|_K^4))^\wedge(\xi), \qquad \xi \in \mathbb{R}^n.$$

First, we show that our choice of the function f guarantees integrability of μ_K for any infinitely smooth body K.

LEMMA 3.19. *Let K be an origin-symmetric infinitely smooth star body. Then the function μ_K is continuous and integrable on $\mathbb{R}^n$.*

PROOF. Let n be an even integer. Consider a function

$$F(x) = \Delta^{(n+2)/2} \exp(-\|x\|_K^4) = h(x) \exp(-\|x\|_K^4), \qquad x \in \mathbb{R}^n,$$

where Δ is the Laplace operator. The function h that appears in front of the exponential after differentiation is the sum of homogeneous functions of degrees $-n+2$ and higher, which are continuous on the sphere S^{n-1}, because K is infinitely smooth (after each differentiation the degree of homogeneity decreases by 1; this explains the choice of the power 4 in the formula for the function f). Clearly, h is locally integrable, and, therefore, F is integrable on $\mathbb{R}^n$. Using the connection between the Fourier transform and differentiation (2.36), we get

$$\hat{F}(x) = (-1)^{(n+2)/2}|x|_2^{n+2}\mu_K(x).$$

The Fourier transform of an integrable function F is a bounded function on $\mathbb{R}^n$. Thus, there exists a constant C so that $|\mu_K(x)| \le C|x|_2^{-n-2}$ for every $x \in \mathbb{R}^n$. The function $|\cdot|_2^{-n-2}$ is integrable outside of the unit ball in $\mathbb{R}^n$. Also, since $\exp(-\|\cdot\|_K^4)$ is integrable on $\mathbb{R}^n$, μ_K is a continuous function and is locally integrable, so the result follows. A similar argument (with $\Delta^{(n+1)/2}$) works in the case where n is an odd integer. □

Let us now prove formula (3.20) in our particular case.

LEMMA 3.20. *Let $0 < p < n$ and let K be an infinitely smooth origin-symmetric star body in $\mathbb{R}^n$. Then for every $\theta \in S^{n-1}$*

$$\frac{\Gamma((n-p)/4)}{4}(\|\cdot\|_K^{-n+p})^\wedge(\theta) = \int_0^\infty t^{p-1}\mu_K(t\theta)\, dt.$$

PROOF. For any even test function $\phi \in \mathcal{S}(\mathbb{R}^n)$, consider the integral

$$\int_{\mathbb{R}^n} \mu_K(x) \left(\int_0^\infty t^{n-p-1}\phi(tx)\, dt \right) dx. \tag{3.23}$$

This integral converges absolutely, because by Lemma 3.19, the function μ_K is continuous and integrable, and the inner integral represents a continuous on $\mathbb{R}^n \setminus \{0\}$ and homogeneous of degree $-n+p$ function of x, so we can apply the argument of Lemma 2.1.

First, let us write the integral (3.23) in spherical coordinates (r, θ) (we also make a substitution $z = rt$ in the inner integral). We get that

$$(3.23) = \int_{S^{n-1}} \left(\int_0^\infty r^{p-1}\mu_K(r\theta)\, dr \right) \left(\int_0^\infty z^{n-p-1}\phi(z\theta)\, dz \right) d\theta.$$

Now let us write the same integral in a different way. Note that, by Corollary 3.17, $(\|\cdot\|_K^{-n+p})^\wedge$ is a continuous function on S^{n-1}. This function is homogeneous of degree $-p$ on $\mathbb{R}^n$. Also, $(\exp(-t^4\|\cdot\|_K^4))^\wedge(\xi) = t^{-n}\mu_K(\xi/t)$. We have

$$\begin{aligned}
(3.23) &= \int_0^\infty t^{n-p-1}\left(\int_{\mathbb{R}^n} \phi(tx)\mu_K(x)\,dx\right)dt \\
&= \int_0^\infty t^{n-p-1}\left(\int_{\mathbb{R}^n} \phi(\xi)\mu_K(\xi/t)t^{-n}\,d\xi\right)dt \\
&= \int_0^\infty t^{n-p-1}\left(\int_{\mathbb{R}^n} \hat{\phi}(y)\exp(-t^4\|y\|_K^4)\,dy\right)dt \\
&= \int_{\mathbb{R}^n} \hat{\phi}(y)\left(\int_0^\infty t^{n-p-1}\exp(-t^4\|y\|_K^4)\,dt\right)dy \\
&= \frac{\Gamma((n-p)/4)}{4}\int_{\mathbb{R}^n} \|y\|_K^{-n+p}\hat{\phi}(y)\,dy \\
&= \frac{\Gamma((n-p)/4)}{4}\langle \|\cdot\|_K^{-n+p}, \hat{\phi}\rangle \\
&= \frac{\Gamma((n-p)/4)}{4}\langle (\|\cdot\|_K^{-n+p})^\wedge, \phi\rangle \\
&= \frac{\Gamma((n-p)/4)}{4}\int_{\mathbb{R}^n} (\|\cdot\|_K^{-n+p})^\wedge(\xi)\phi(\xi)\,d\xi \\
&= \frac{\Gamma((n-p)/4)}{4}\int_{S^{n-1}} (\|\cdot\|_K^{-n+p})^\wedge(\theta)\left(\int_0^\infty r^{n-p-1}\phi(r\theta)\,dr\right)d\theta.
\end{aligned}$$

Now if we put $\phi(x) = u(r)v(\theta)$ in both expressions for the integral (3.23), where v is any infinitely differentiable function on the sphere and u is a non-negative even test function on $\mathbb{R}$ with compact support outside of zero, we get

$$\begin{aligned}
&\int_{S^{n-1}}\left(\int_0^\infty r^{p-1}\mu_K(r\theta)\,dr\right)v(\theta)\,d\theta \\
&= \frac{\Gamma((n-p)/4)}{4}\int_{S^{n-1}} (\|\cdot\|_K^{-n+p})^\wedge(\theta)v(\theta)\,d\theta
\end{aligned}$$

for every $v \in C^\infty(S^{n-1})$. The result follows. □

We need to apply Parseval's formula on $\mathbb{R}^n$ in the case where one of the functions does not belong to $L_1(\mathbb{R}^n)$ or $L_2(\mathbb{R}^n)$.

LEMMA 3.21. *Let $0 < p < n$, and let K and D be infinitely smooth origin-symmetric star bodies in $\mathbb{R}^n$. Then*

$$(3.24)\qquad (2\pi)^n\int_{\mathbb{R}^n} \|x\|_K^{-p}\exp(-\|x\|_D^4)\,dx = \int_{\mathbb{R}^n} (\|\cdot\|_K^{-p})^\wedge(\xi)\,\mu_D(\xi)\,d\xi.$$

PROOF. By Lemma 2.1, the integrals in both sides of (3.24) converge absolutely. By Lemma 3.20 (with p in place of $n-p$; we extend both sides of the equation to the whole $\mathbb{R}^n$ as homogeneous functions of degree $-n+p$), the right-hand side of (3.24) is equal to

$$\frac{4}{\Gamma(p/4)}\int_{\mathbb{R}^n}\left(\int_0^\infty t^{n-p-1}\mu_K(t\xi)\,dt\right)\mu_D(\xi)\,d\xi.$$

By the Fubini theorem, the latter is equal to

$$\frac{4}{\Gamma(p/4)}\int_0^\infty t^{n-p-1}\left(\int_{\mathbb{R}^n}\mu_K(t\xi)\mu_D(\xi)\,d\xi\right)\,dt.$$

We can apply Parseval's formula for L_1-functions (see (2.34)) to the inner integral, because μ_K and μ_D are even continuous integrable functions on $\mathbb{R}^n$, which are the Fourier transforms of the functions $\exp(-\|\cdot\|_K^4)$ and $\exp(-\|\cdot\|_D^4)$. We continue the calculation:

$$= \frac{4(2\pi)^n}{\Gamma(p/4)}\int_0^\infty t^{-p-1}\left(\int_{\mathbb{R}^n}\exp(-\|x\|_K^4/t^4)\exp(-\|x\|_D^4)\,dx\right)\,dt$$

$$= \frac{4(2\pi)^n}{\Gamma(p/4)}\int_{\mathbb{R}^n}\left(\int_0^\infty t^{-p-1}\exp(-\|x\|_K^4/t^4)\,dt\right)\exp(-\|x\|_D^4)\,dx$$

$$= (2\pi)^n\int_{\mathbb{R}^n}\|x\|_K^{-p}\exp(-\|x\|_D^4)\,dx.$$

□

We are now ready to prove Parseval's formula on the sphere in the infinitely smooth case.

LEMMA 3.22. *Let K and D be infinitely smooth origin-symmetric star bodies in $\mathbb{R}^n$, and let $0<p<n$. Then*

$$\int_{S^{n-1}}(\|\cdot\|_K^{-p})^\wedge(\theta)(\|\cdot\|_D^{-n+p})^\wedge(\theta)\,d\theta = (2\pi)^n\int_{S^{n-1}}\|\theta\|_K^{-p}\|\theta\|_D^{-n+p}\,d\theta.$$

PROOF. Passing to spherical coordinates, we get

$$\int_{\mathbb{R}^n}\|x\|_K^{-p}\exp(-\|x\|_D^4)\,dx \tag{3.25}$$

$$= \int_{S^{n-1}}\|\theta\|_K^{-p}\left(\int_0^\infty r^{n-p-1}\exp(-r^4\|\theta\|_D^4)\,dr\right)\,d\theta$$

$$= \frac{\Gamma((n-p)/4)}{4}\int_{S^{n-1}}\|\theta\|_K^{-p}\|\theta\|_D^{-n+p}\,d\theta.$$

Now use Lemma 3.21 to get a different expression for the integral in (3.25):

$$\int_{\mathbb{R}^n}\|x\|_K^{-p}\exp(-\|x\|_D^4)\,dx$$

$$= \frac{1}{(2\pi)^n}\int_{\mathbb{R}^n}(\|\cdot\|_K^{-p})^\wedge(\xi)\mu_D(\xi)\,d\xi$$

$$= \frac{1}{(2\pi)^n}\int_{S^{n-1}}(\|\cdot\|_K^{-p})^\wedge(\theta)\left(\int_0^\infty r^{p-1}\mu_D(r\theta)\,dr\right)\,d\theta.$$

The result follows from Lemma 3.20. □

We can use an approximation argument to weaken the smoothness assumptions on the bodies K and D. The coefficient $(2\pi)^n$ disappears because of our definition of the measure μ_0 in Corollary 2.26, which is more convenient in other applications.

COROLLARY 3.23. *Let $k \in \mathbb{N}$, $1 \le k \le n-1$, and let K and D be origin-symmetric star bodies in $\mathbb{R}^n$. Suppose that D is $(k-1)$-smooth if k is odd and that D is k-smooth if k is even. Also suppose that $\|\cdot\|_K^{-k}$ is a positive definite distribution, and let μ_0 be the finite Borel measure on S^{n-1} that corresponds to $\|\cdot\|_K^{-k}$ by Corollary 2.26. Then*

$$\int_{S^{n-1}} (\|\cdot\|_D^{-n+k})^\wedge(\theta)\, d\mu_0(\theta) = \int_{S^{n-1}} \|\theta\|_K^{-k}\|\theta\|_D^{-n+k}\, d\theta. \tag{3.26}$$

PROOF. Since D is $(k-1)$-smooth, we can approximate D by a sequence D_m of infinitely smooth symmetric star bodies in such a way that the functions $\|\cdot\|_{D_m}$ converge to $\|\cdot\|_D$ in the metric of the space $C^{k-1}(S^{n-1})$; see the end of Section 2.2. Then, by Corollary 3.17, the Fourier transforms $(\|\cdot\|_{D_m}^{-n+k})^\wedge$ are continuous functions on the sphere that converge in $C(S^{n-1})$ to another continuous function $(\|\cdot\|_D^{-n+k})^\wedge$, as $m \to \infty$.

Also consider any sequence of infinitely smooth symmetric star bodies $K_j \subset K$, $j \in \mathbb{N}$, approximating K in the radial metric. For every even test function ϕ the sequence of integrable (by Lemma 2.1) functions $\|\cdot\|_{K_j}^{-k}|\hat{\phi}(\cdot)|$ is majorated by an integrable function $\|\cdot\|_K^{-k}|\hat{\phi}(\cdot)|$. By the dominated convergence theorem and the first part of Corollary 2.26,

$$\int_{\mathbb{R}^n} (\|\cdot\|_{K_j}^{-k})^\wedge(y)\phi(y)\, dy = \int_{\mathbb{R}^n} \|x\|_{K_j}^{-k}\hat{\phi}(x)\, dx$$

$$\to \int_{\mathbb{R}^n} \|x\|_K^{-k}\hat{\phi}(x)\, dx = (2\pi)^n \int_{S^{n-1}} \left(\int_0^\infty r^{k-1}\phi(r\theta)\, dr\right) d\mu_0(\theta), \tag{3.27}$$

as $j \to \infty$. Since the bodies D_m are infinitely smooth,

$$\phi_m(x) = u(r)(\|\cdot\|_{D_m}^{-n+k})^\wedge(\theta)$$

is a test function, where u is any non-negative even test function on $\mathbb{R}$ with compact support outside of zero. Substituting ϕ_m in (3.27) and passing to spherical coordinates, we get

$$\lim_{j\to\infty} \int_{S^{n-1}} (\|\cdot\|_{K_j}^{-k})^\wedge(\theta)(\|\cdot\|_{D_m}^{-n+k})^\wedge(\theta)\, d\theta$$

$$= (2\pi)^n \int_{S^{n-1}} (\|x\|_{D_m}^{-n+k})^\wedge(\theta)\, d\mu_0(\theta). \tag{3.28}$$

By Corollary 3.17, the functions $(\|\cdot\|_{D_m}^{-n+k})^\wedge(\theta)$ converge in $C(S^{n-1})$ to the function $(\|\cdot\|_D^{-n+k})^\wedge(\theta)$, as $m \to \infty$, so the integral in the right-hand side of (3.28) converges to the integral in the left-hand side of (3.26). Now we get the result if we apply Lemma 3.22 to the integral in the left-hand side of (3.28) and let $m, j \to \infty$. □

The technique of this section allows us to prove a spherical version of another basic fact from Fourier analysis.

LEMMA 3.24. *Let $1 \le k < n$, let ϕ be an even continuous integrable function on $\mathbb{R}^n$, and let H be an $(n-k)$-dimensional subspace of $\mathbb{R}^n$. Suppose that ϕ is integrable on all translations of H and that the Fourier transform $\hat{\phi}$ is integrable on $H^\perp$. Then*

$$(2\pi)^k \int_H \phi(x)\, dx = \int_{H^\perp} \hat{\phi}(x)\, dx.$$

PROOF. Let $\xi_1, ..., \xi_k$ be an orthonormal basis in $H^\perp$. For every $t \in \mathbb{R}^k$, making the change of variables $u_1 = (x, \xi_1), ..., u_k = (x, \xi_k)$, we get

$$\hat{\phi}(t_1\xi_1 + ... + t_k\xi_k) = \int_{\mathbb{R}^n} \phi(x) \exp(-i(x, t_1\xi_1 + ... + t_k\xi_k))\, dx$$

$$= \int_{\mathbb{R}^n} \phi(x) \exp(-it_1(x, \xi_1) + ... + t_k(x, \xi_k))\, dx$$

$$= \int_{\mathbb{R}^k} \exp(-i(t_1u_1 + ... + t_ku_k)) \left(\int_{H+u_1\xi_1+...+u_k\xi_k} \phi(x)\, dx \right) du.$$

This means that the function $f(t) = \hat{\phi}(t_1\xi_1 + ... + t_k\xi_k)$ is the Fourier transform of the function $g(u) = \int_{H+u_1\xi_1+...+u_k\xi_k} \phi(x)\, dx$. The function g is even integrable on $\mathbb{R}^k$ because ϕ is even integrable on $\mathbb{R}^n$. Also, since $\hat{\phi}$ is even integrable on $H^\perp$, the function f is integrable and even on $\mathbb{R}^k$. By the inversion theorem for the Fourier transform, [Ru, Th. 7.7],

$$\int_{\mathbb{R}^k} f(t)\, dt = (\hat{g})^\wedge(0) = (2\pi)^k g(0),$$

which immediately implies the result, because the left-hand side is exactly $\int_{H^\perp} \hat{\phi}$. □

The following spherical version of Lemma 3.24 provides a Fourier transform formula for the volume of lower dimensional sections. This formula extends the result of Theorem 3.8 and will be useful in several places later.

LEMMA 3.25. *Suppose that $1 \le k < n$ is an integer and that L is an origin-symmetric star body in $\mathbb{R}^n$ so that L is $(k-1)$-smooth if k is odd and it is k-smooth if k is even. Then for every $(n-k)$-dimensional subspace H of $\mathbb{R}^n$ we have*

$$\mathrm{Vol}_{n-k}(L \cap H) = \frac{1}{(2\pi)^k(n-k)} \int_{S^{n-1}\cap H^\perp} (\|\cdot\|_L^{-n+k})^\wedge(\theta)\, d\theta.$$

PROOF. We first assume that L is infinitely smooth. By Lemma 3.20,

$$\int_{S^{n-1}\cap H^\perp} (\|\cdot\|_L^{-n+k})^\wedge(\theta)\, d\theta$$

$$= (4/\Gamma((n-k)/4)) \int_{S^{n-1}\cap H^\perp} \left(\int_0^\infty t^{k-1} \mu_L(t\theta)\, dt \right) d\theta$$

$$= (4/\Gamma((n-k)/4)) \int_{H^\perp} \mu_L(x)\, dx.$$

Using Lemma 3.24 with $\phi = \exp(-\|\cdot\|_L^4)$ and writing the integral in polar coordinates, we get

$$= (4(2\pi)^k/\Gamma((n-k)/4)) \int_H \exp(-\|x\|_L^4)\, dx$$

$$= (2\pi)^k \int_{S^{n-1}\cap H} \|\theta\|_L^{-n+k}\, d\theta = (2\pi)^k(n-k)\mathrm{Vol}_{n-k}(L \cap H),$$

by (2.3). The assumption that L is infinitely smooth can be removed by approximation in the same way, as was done in Corollary 3.23. □

3.5. Remarks and further results

As mentioned in the Introduction, the case of Theorem 3.1 where ξ is parallel to the main diagonal of the cube was known to Laplace [La]. To the best of the author's knowledge, the general formula of Theorem 3.1 first appeared in the paper [Po1] by Polya. This formula was independently derived and applied to the study of hyperplane sections of the cube by Hensley [He1] and Ball [Ba1]. The proof here is the same as in the paper [NP] by Nazarov and Podkorytov.

The formula of Theorem 3.2 in the case $1 \le q \le 2$ was found by Meyer and Pajor [MeP, Lemma II.6]. The proof uses stable random variables. The validity of this formula for other values of $q \in (0,\infty)$ was established in [K7]. The formulas for the Fourier transform of ℓ_q^n-norms, $0 < q \le \infty$, from Lemmas 3.4 and 3.6 are taken from [K8].

The general Fourier transform formula for the volume of hyperplane sections of star bodies, Theorem 3.8, was proved in [K7]. Together with the results of Lemmas 3.4 and 3.6, this formula implies Theorem 3.1 and proves Theorem 3.2 for all $0 < q < \infty$.

The result of Lemma 3.16 represents a way to invert the spherical Radon transform and cosine transforms with negative exponents. This approach was suggested in [K6]. As the author has recently learned, similar formulas relating the Radon and cosine transforms of homogeneous functions to the Fourier transform were proved by Semyanistyi [Se]. A geometric way of inverting the spherical Radon transform in certain situations was used by Gardner [Ga2] and Zhang [Zh2] in their solutions of the Busemann-Petty problem in dimensions 3 and 4. One can find more results on the connection between the Radon, Fourier and cosine transforms; related works include [Al], [AB], [GW1], [Gri], [Rb1], [Rb2], [SW].

Theorem 3.18 was proved in [GKS]. This result contains Theorem 3.8 as a particular case (with $k = 0$) and plays the key role in the unified solution to the Busemann-Petty problem; see Section 5.1. A way to prove similar formulas without using the Fourier transform or analytic continuation was suggested by Barthe, Fradelizi and Maurey in [BFM].

The result of Theorem 3.18 was extended to sections of lower dimensions in [K14]. Let $H \in G(n, n-k)$, $1 \le k < n$, and let $\xi_1, ..., \xi_k$ be an orthonormal basis in $H^\perp$. For a convex body K in $\mathbb{R}^n$, the $(n-k)$-dimensional parallel section function $A_{K,H}$ is a function on $\mathbb{R}^k$ defined by

$$A_{K,H}(u) = \mathrm{Vol}_{n-k}(K \cap \{H + u_1\xi_1 + ... + u_k\xi_k\}),\ u \in \mathbb{R}^k.$$

The result in [K14] is as follows:

THEOREM 3.26. *Let K be an infinitely smooth origin-symmetric convex body in $\mathbb{R}^n$, $1 \le k < n$. Then for every $(n-k)$-dimensional subspace H of $\mathbb{R}^n$ and every $m \in \mathbb{N} \cup \{0\}$, $m \ne (n-k)/2$,*

$$\Delta^m A_{K,H}(0) = \frac{(-1)^m}{2^k \pi^k (n-2m-k)} \int_{S^{n-1}\cap H^\perp} (\|x\|_K^{-n+2m+k})^\wedge(\theta)\, d\theta.$$

The case $m = 0$ of Theorem 3.26 provides an alternative proof of Lemma 3.25. The proof of Theorem 3.26 follows the lines of the proof of Theorem 3.18.

The spherical Parseval formulas of Lemma 3.22 and Corollary 3.23 were proved in [K13]. These formulas are used in many places, most importantly in the generalizations of the Busemann-Petty problem in Sections 5.2 and 5.4. We presented here the original proof from [K13]. This proof projects the classical Parseval formula to the sphere. An alternative proof using spherical harmonics is shorter (see Remark (ii) at the end of Section 5.3), but it does not give a clear idea of what is going on. The spherical Parseval formula (with $p = 1$) combined with Theorem 3.8 gives an expression for the volume of a body in terms of volumes of central sections: for any $(n-2)$-smooth origin-symmetric convex body D in $\mathbb{R}^n$,

$$\mathrm{Vol}_n(D) = \frac{(2\pi)^n \pi(n-1)}{n} \int_{S^{n-1}} \mathrm{Vol}_{n-1}(D \cap \xi^\perp)(\|\cdot\|_D^{-1})^\wedge(\xi)\, d\xi.$$

The formula of Lemma 3.25 was first proved in [K13]. This formula is used in Section 4.2 to characterize k-intersection bodies.

CHAPTER 4

Intersection Bodies

Let K, L be origin-symmetric star bodies in $\mathbb{R}^n$. Following Lutwak [Lu1], we say that K is the intersection body of L if the radius of K in every direction is equal to the $(n-1)$-dimensional volume of the section of L by the central hyperplane orthogonal to this direction, i.e. for every $\xi \in S^{n-1}$,

$$\rho_K(\xi) = \|\xi\|_K^{-1} = \mathrm{Vol}_{n-1}(L \cap \xi^\perp). \tag{4.1}$$

All the bodies K that appear as intersection bodies of different star bodies form *the class of intersection bodies of star bodies.* It follows from Corollary 3.9 that if two origin-symmetric star bodies L_1, L_2 generate the same intersection body, then $L_1 = L_2$.

Note that the right-hand side of (4.1) can be written using (2.2):

$$\|\xi\|_K^{-1} = \frac{1}{n-1}\int_{S^{n-1}\cap\xi^\perp} \|\theta\|_L^{-n+1} d\theta = \frac{1}{n-1} R(\|\cdot\|_L^{-n+1})(\xi),$$

where R is the spherical Radon transform. It means that a star body K is the intersection body of a star body if and only if the function $\|\cdot\|_K^{-1}$ is the spherical Radon transform of a continuous positive function on S^{n-1}. This allows us to introduce a more general class of bodies. We say that a star body K in $\mathbb{R}^n$ is an *intersection body* if there exists a finite Borel measure μ on the sphere S^{n-1} so that $\|\cdot\|_K^{-1} = R\mu$ as functionals on $C(S^{n-1})$, i.e. for every continuous function f on S^{n-1},

$$\int_{S^{n-1}} \|x\|_K^{-1} f(x)\, dx = \int_{S^{n-1}} Rf(x)\, d\mu(x). \tag{4.2}$$

In this chapter we study the class of intersection bodies and its generalizations.

4.1. A Fourier analytic characterization

We are going to prove in this section that an origin-symmetric star body K in $\mathbb{R}^n$ is an intersection body if and only if the function $\|\cdot\|_K^{-1}$ represents a positive definite distribution. We shall see later that this characterization is quite effective in the study of intersection bodies. Let us start with a simple argument showing a particular case of this result.

Suppose K is the intersection body of some symmetric star body L. Then, using Theorem 3.8, for every $x \in S^{n-1}$, we get

$$\|x\|_K^{-1} = \mathrm{Vol}_{n-1}(L \cap x^\perp) = \frac{1}{\pi(n-1)}(\|\cdot\|_L^{-n+1})^\wedge(x).$$

We extend this equality to the whole $\mathbb{R}^n$, as an equality of homogeneous functions of degree -1. Since K is symmetric, the Fourier transform is self-invertible (up to a constant), so

$$(\|\cdot\|_K^{-1})^\wedge(\xi) = \frac{(2\pi)^n}{\pi(n-1)}\|\xi\|_L^{-n+1} > 0,$$

which means that the locally integrable function $\|\cdot\|_K^{-1}$ represents a positive definite distribution.

On the other hand, suppose that $\|\cdot\|_K^{-1}$ is positive definite and its Fourier transform is a continuous positive function on S^{n-1}. Then define a symmetric star body L by

$$\|\xi\|_L = \left(\frac{\pi(n-1)}{(2\pi)^n}(\|\cdot\|_K^{-1})^\wedge(\xi)\right)^{-1/(n-1)}, \qquad \xi \in S^{n-1}.$$

It is easy to reverse the argument above and to see that K is the intersection body of L. We have proved that an origin-symmetric star body K in $\mathbb{R}^n$ is an intersection body of a star body if and only if the Fourier transform of $\|\cdot\|_K^{-1}$ is a continuous positive function on the sphere.

A similar Fourier analytic characterization holds for the whole class of intersection bodies. The argument is essentially the same, but it looks more complicated because we have to deal with measures.

THEOREM 4.1. *An origin-symmetric star body K in $\mathbb{R}^n$ is an intersection body if and only if $\|\cdot\|_K^{-1}$ represents a positive definite distribution on $\mathbb{R}^n$.*

PROOF. By Lemma 2.11, for every even test function ϕ and every fixed $\theta \in S^{n-1}$, the function $f(t) = (2\pi)^n\phi(t\theta)$ is the Fourier transform of the function $g(z) = \mathcal{R}\hat{\phi}(\theta; z)$, where $t, z \in \mathbb{R}$ and $\mathcal{R}$ is the Radon transform. Inverting the Fourier transform, we get $\int_{\mathbb{R}} f = 2\pi g(0)$, which can be written as

$$(2\pi)^n \int_{\mathbb{R}} \phi(t\theta)\, dt = 2\pi\mathcal{R}\hat{\phi}(\theta; 0) = 2\pi \int_{\theta^\perp} \hat{\phi}(x)\, dx.$$

Writing the integral in the right-hand side in the polar coordinates of the hyperplane $\theta^\perp$, we get

$$2\pi \int_{S^{n-1}\cap\theta^\perp} \left(\int_0^\infty t^{n-2}\hat{\phi}(t\xi)\, dt\right) d\xi = 2\pi R\left(\int_0^\infty t^{n-2}\hat{\phi}(t\xi)\, dt\right)(\theta),$$

where in the right-hand side we have the spherical Radon transform of a function of the variable $\xi \in S^{n-1}$ at the point θ. Combining the last two formulas, we get

$$(2\pi)^{n-1} \int_{\mathbb{R}} \phi(t\theta)\, dt = R\left(\int_0^\infty t^{n-2}\hat{\phi}(t\xi)\, dt\right)(\theta). \tag{4.3}$$

Suppose that K is an intersection body, and let μ be the measure on S^{n-1} for which $R\mu = \|\cdot\|_K^{-1}$ in the sense of equality (4.2). By Lemma 2.1, $\|\cdot\|_K^{-1}$ is a locally integrable function on $\mathbb{R}^n$, so, for every non-negative even test function ϕ, we have

$$\langle(\|\cdot\|_K^{-1})^\wedge, \phi\rangle = \langle\|\cdot\|_K^{-1}, \hat{\phi}\rangle = \int_{\mathbb{R}^n} \|x\|_K^{-1}\hat{\phi}(x)\, dx.$$

We need to prove that the latter integral is non-negative. Writing this integral in polar coordinates, we get that it is equal to

$$\int_{S^{n-1}} \|\xi\|_K^{-1} \left(\int_0^\infty t^{n-2}\hat{\phi}(t\xi)\, dt\right) d\xi,$$

which, by (4.2), equals

$$\int_{S^{n-1}} R\left(\int_0^\infty t^{n-2}\hat{\phi}(t\xi)\ dt\right)(\theta)\ d\mu(\theta).$$

Applying (4.3), we write the latter integral as

$$(2\pi)^{n-1}\int_{S^{n-1}}\left(\int_{\mathbb{R}} \phi(t\theta)\ dt\right)\ d\mu(\theta) \geq 0,$$

because ϕ is a non-negative test function. Therefore, $\|\cdot\|_K^{-1}$ represents a positive definite distribution.

Conversely, suppose that $\|\cdot\|_K^{-1}$ represents a positive definite distribution. By Corollary 2.26, there exists a finite Borel measure μ_0 on S^{n-1} so that for every even test function ϕ

$$\langle \|\cdot\|_K^{-1}, \phi\rangle = \int_{S^{n-1}}\left(\int_{\mathbb{R}} \hat{\phi}(t\theta)\ dt\right) d\mu_0(\theta).$$

Now, by (4.3) with $\hat{\phi}$ in place of ϕ, the latter expression is equal to

$$2\pi\int_{S^{n-1}} R\left(\int_0^\infty t^{n-2}\phi(t\xi)\ dt\right)(\theta)\ d\mu_0(\theta). \tag{4.4}$$

On the other hand,

$$\langle \|\cdot\|_K^{-1}, \phi\rangle = \int_{S^{n-1}} \|\theta\|_K^{-1}\left(\int_0^\infty t^{n-2}\phi(t\theta)\ dt\right) d\theta. \tag{4.5}$$

For any infinitely differentiable function v on the sphere S^{n-1}, consider a test function $\phi \in \mathcal{S}(\mathbb{R}^n)$ defined by $\phi(x) = u(t)v(\theta)$ for every $x = t\theta \in \mathbb{R}^n,\ t > 0,\ \theta \in S^{n-1}$, where $u \in \mathcal{S}(\mathbb{R})$ is a non-negative even function with compact support outside of zero and $\int_{\mathbb{R}} t^{n-2}u(t)dt = 1$. We have

$$v(\theta) = \int_0^\infty t^{n-2}\phi(t\theta)\ dt.$$

Substituting this ϕ in (4.4) and (4.5), we get that for any $v \in C^\infty(S^{n-1})$,

$$\int_{S^{n-1}} \|\theta\|_K^{-1}v(\theta)\ d\theta = 2\pi\int_{S^{n-1}} Rv(\theta)\ d\mu_0(\theta).$$

Since infinitely differentiable functions are dense in $C(S^{n-1})$, the latter equality holds for any continuous function v on the sphere, so, by (4.2), the functionals $\|\cdot\|_K^{-1}$ and $R(2\pi\mu_0)$ are equal on $C(S^{n-1})$, which means that K is an intersection body. □

The result of Theorem 4.1 allows us to give many examples of bodies that are and that are not intersection bodies. Most of these examples will be presented in Sections 4.3 and 4.4 in a more general setting of k-intersection bodies. In this section we include two applications that later will be used in the solution of the Busemann-Petty problem.

COROLLARY 4.2. *Every origin-symmetric convex body in* $\mathbb{R}^4$ *is an intersection body.*

PROOF. In view of Lemma 4.10 (see the end of Section 4.2), it is enough to prove that every origin-symmetric infinitely smooth convex body K in $\mathbb{R}^4$ is an intersection body. To do this, first note that, by Corollary 3.17, the Fourier transform of $\|\cdot\|_K^{-1}$ is a homogeneous of degree $-n+1$ continuous on $\mathbb{R}^n \setminus \{0\}$ function. By Theorem 3.18 with $n=4$, $k=2$, for every $\xi \in S^{n-1}$,

$$(\|\cdot\|_K^{-1})^\wedge(\xi) = -\pi A''_{K,\xi}(0).$$

Since the body K is convex, by Brunn's theorem, Theorem 2.3, the quantity in the right-hand side of the latter equality is non-negative, so the function $(\|\cdot\|_K^{-1})^\wedge$ is non-negative on S^{n-1}. This implies that $(\|\cdot\|_K^{-1})^\wedge$ is a positive distribution, and, by Theorem 4.1, K is an intersection body. □

The result of Corollary 4.2 is no longer true in dimensions 5 and higher. In order to explain this transition between the dimensions 4 and 5, let us look at a geometric characterization of intersection bodies that is an immediate consequence of Theorems 3.18 and 4.1.

COROLLARY 4.3. *If n is an even integer, an origin-symmetric infinitely smooth convex body K in $\mathbb{R}^n$ is an intersection body if and only if for every $\xi \in S^{n-1}$,*

$$\xi \mapsto (-1)^{(n-2)/2} A_{K,\xi}^{(n-2)}(0)$$

is a non-negative function on S^{n-1}.

If n is an odd integer, then K is an intersection body if and only if for every $\xi \in S^{n-1}$,

$$(-1)^{(n-1)/2} \int_0^\infty \frac{A_{K,\xi}(z) - A_{K,\xi}(0) - \dots - A_{K,\xi}^{(n-3)}(0)\frac{z^{n-3}}{(n-3)!}}{z^{n-1}}\, dz \geq 0.$$

Put $n=4$ in the latter statement and suppose, in addition, that K is convex. By Brunn's theorem, as in Corollary 4.2, K is an intersection body. If $n=5$, we are essentially looking at the third derivative of the parallel section function. So the transition between dimensions 4 and 5 is a result of the fact that convexity controls the second derivatives, but it does not control the third and higher order derivatives. The following example illustrates this argument.

COROLLARY 4.4. *There exists an origin-symmetric convex body K in $\mathbb{R}^5$ that is not an intersection body.*

PROOF. Put $n=5$ in Corollary 4.3. To get an example of a non-intersection body in dimension 5, we need to construct an origin-symmetric infinitely smooth convex body K in $\mathbb{R}^5$ for which there exists a $\xi \in S^4$ such that the parallel section function $A_{K,\xi}$ of K satisfies

$$\int_0^\infty \frac{A_{K,\xi}(z) - A_{K,\xi}(0) - A''_{K,\xi}(0)\frac{z^2}{2}}{z^4}\, dz < 0. \tag{4.6}$$

Let $N>0$, define $f_N(x) = (1-x^2-Nx^4)^{1/4}$, and let $a_N>0$ be such that $f_N(a_N)=0$ and $1-x^2-Nx^4>0$ on $(0,a_N)$. Clearly,

$$\lim_{N\to\infty} N^{1/4} a_N = 1. \tag{4.7}$$

The function f_N has its maximum at 0 and

$$f''_N(x) = -(\frac{1}{2} + 3Nx^2)(1-x^2-Nx^4)^{-3/4}$$

$$-3(\frac{1}{2}x + Nx^3)^2(1 - x^2 - Nx^4)^{-7/4},$$

so f_N is strictly concave on $[0, a_N]$. It follows that

$$K = \left\{ (x_1, \ldots, x_5) \in \mathbb{R}^5 : x_5 \in [-a_N, a_N], \left(\sum_{i=1}^{4} x_i^2 \right)^{1/2} \leq f_N(|x_5|) \right\}$$

is a strictly convex infinitely smooth body. For $0 \leq z \leq a_N$,

$$K \cap \{(x_1, \ldots, x_5) \in \mathbb{R}^5 : x_5 = z\}$$

is a four-dimensional Euclidean ball of radius $f_N(z)$. If $\xi = (0, 0, 0, 0, 1)$, then

$$A_{K,\xi}(z) = \frac{\pi^2}{2} f_N(z)^4 = \frac{\pi^2}{2}(1 - z^2 - Nz^4), \quad 0 \leq z \leq a_N.$$

This implies that, for our choice of ξ, the integral (4.6) equals

$$\frac{\pi^2}{2}(-Na_N + \frac{1}{a_N} - \frac{1}{3a_N^3}),$$

which, by (4.7), becomes negative when N is large enough. □

4.2. k-intersection bodies

Let $1 \leq k < n$, and let D and L be origin-symmetric star bodies in $\mathbb{R}^n$. Following [K13], we say that D is a *k-intersection body of L* if for every $(n-k)$-dimensional subspace H of $\mathbb{R}^n$

$$\text{Vol}_k(D \cap H^\perp) = \text{Vol}_{n-k}(L \cap H). \tag{4.8}$$

Unlike the case of intersection bodies (corresponding to $k = 1$ in the latter definition, up to a constant 2), for $k > 1$ it is not immediately clear when a k-intersection body exists for a given L.

By the polar formula for the volume (2.3), equality (4.8) can be written in the form

$$\int_{S^{n-1} \cap H^\perp} \|\xi\|_D^{-k} d\xi = \frac{k}{n-k} \int_{S^{n-1} \cap H} \|x\|_L^{-n+k} dx. \tag{4.9}$$

Let us first consider the case where L is infinitely smooth. Then, by Corollary 3.17, $(\|\cdot\|_L^{-n+k})^\wedge$ is a continuous function on the sphere. Moreover, we can apply Lemma 3.25 to the right-hand side of (4.9): for every $(n-k)$-dimensional subspace H of $\mathbb{R}^n$,

$$\int_{S^{n-1} \cap H^\perp} \|\xi\|_D^{-k} \, d\xi = \frac{k}{(n-k)(2\pi)^k} \int_{S^{n-1} \cap H^\perp} (\|\cdot\|_L^{-n+k})^\wedge(\xi) \, d\xi.$$

Now, by Corollary 3.10, the two functions of the variable ξ are equal everywhere on S^{n-1}, and we get the following lemma.

LEMMA 4.5. *Let L be an infinitely smooth origin-symmetric star body in $\mathbb{R}^n$, and let $1 \leq k < n$. The k-intersection body D of L exists if and only if the function $(\|\cdot\|_L^{-n+k})^\wedge$ is strictly positive on S^{n-1}. The body D is then given by*

$$\|\xi\|_D^{-k} = \frac{k}{(n-k)(2\pi)^k}(\|\cdot\|_L^{-n+k})^\wedge(\xi)$$

for every $\xi \in S^{n-1}$.

On the other hand, reversing the Fourier transform in the latter formula, one can see that if a body D is the k-intersection body of some infinitely smooth symmetric star body, then $(\|\cdot\|_D^{-k})^\wedge$ is a strictly positive continuous function on the sphere. We are going to prove a more general fact by removing the smoothness assumption.

THEOREM 4.6. *Let $1 \le k < n$, and suppose that D is an origin-symmetric star body in $\mathbb{R}^n$ that is the k-intersection body of an origin-symmetric star body L. Then $(\|\cdot\|_D^{-k})^\wedge$ is a strictly positive continuous function on $\mathbb{R}^n \setminus \{0\}$. The corresponding body L satisfies*

$$\|\theta\|_L^{-n+k} = \frac{n-k}{(2\pi)^{n-k}k}(\|\cdot\|_D^{-k})^\wedge(\theta), \ \theta \in S^{n-1}.$$

PROOF. Let $D_m \subset D$, $m \in \mathbb{N}$, be infinitely smooth origin-symmetric star bodies in $\mathbb{R}^n$ approximating D in the radial metric. Then, as $m \to \infty$,

$$\int_{S^{n-1}\cap H^\perp} \|x\|_{D_m}^{-k}\, dx \to \int_{S^{n-1}\cap H^\perp} \|x\|_D^{-k}\, dx$$

uniformly with respect to $H \in G(n, n-k)$. Now we apply the definition of a k-intersection body of a star body in the form of (4.9) to the right-hand side of the latter formula, and we apply Lemma 3.25 to the left-hand side of this formula. We get that

$$\int_{S^{n-1}\cap H} (\|\cdot\|_{D_m}^{-k})^\wedge(\theta)\, d\theta \to \frac{(2\pi)^{n-k}k}{n-k} \int_{S^{n-1}\cap H} \|x\|_L^{-n+k}\, dx \tag{4.10}$$

uniformly with respect to H. Denote the constant in the right-hand side by c. Let $\psi \in C^\infty(S^{n-1})$ be an even infinitely differentiable function on the sphere. Then, by Lemma 3.16, $(\psi(\theta)r^{-1})^\wedge$ is an infinitely differentiable function on $\mathbb{R}^n \setminus \{0\}$. The uniform convergence in (4.10) implies that

$$\lim_{m\to\infty} \int_{G(n,n-k)} \left(\int_{S^{n-1}\cap H} (\|\cdot\|_{D_m}^{-k})^\wedge\right) \left(\int_{S^{n-1}\cap H^\perp} (\psi(\theta)r^{-1})^\wedge\right) dH$$

$$= c\int_{G(n,n-k)} \left(\int_{S^{n-1}\cap H} \|\cdot\|_L^{-n+k}\right) \left(\int_{S^{n-1}\cap H^\perp} (\psi(\theta)r^{-1})^\wedge\right) dH.$$

Applying Lemma 2.12 to both sides of the latter equality, we get

$$\lim_{m\to\infty} \int_{S^{n-1}} (\|\cdot\|_{D_m}^{-k})^\wedge(\xi) \left(\int_{S^{n-1}\cap\xi^\perp} (\psi(\theta)r^{-1})^\wedge\right) d\xi$$

$$= c\int_{S^{n-1}} \|\xi\|_L^{-n+k} \left(\int_{S^{n-1}\cap\xi^\perp} (\psi(\theta)r^{-1})^\wedge\right) d\xi.$$

Since the function $(\psi(\theta)r^{-1})^\wedge$ is homogeneous of degree $-n+1$, by Lemma 3.7, for every $\xi \in S^{n-1}$,

$$\int_{S^{n-1}\cap\xi^\perp} (\psi(\theta)r^{-1})^\wedge = \frac{1}{\pi}\left((\psi(\theta)r^{-1})^\wedge\right)^\wedge(\xi) = \frac{(2\pi)^n}{\pi}\psi(\xi).$$

Therefore,

$$\int_{S^{n-1}} (\|\cdot\|_{D_m}^{-k})^\wedge(\xi)\psi(\xi)\, d\xi \to c\int_{S^{n-1}} \|\xi\|_L^{-n+k}\psi(\xi)\, d\xi, \tag{4.11}$$

as $m \to \infty$.

Let $\phi \in \mathcal{S}(\mathbb{R}^n)$ be any even test function. By Lemma 3.11 and formula (4.11) applied to the function $\psi(\xi) = \int_0^\infty t^{k-1}\phi(t\xi)\, dt$,

$$\langle (\|\cdot\|_D^{-k})^\wedge, \phi\rangle = \int_{\mathbb{R}^n} \|x\|_D^{-k}\hat{\phi}(x)\, dx$$

$$= \lim_{m\to\infty} \int_{\mathbb{R}^n} \|x\|_{D_m}^{-k}\hat{\phi}(x)\, dx = \lim_{m\to\infty} \int_{\mathbb{R}^n} (\|\cdot\|_{D_m}^{-k})^\wedge(y)\phi(y)\, dy$$

$$= \lim_{m\to\infty} \int_{S^{n-1}} (\|\cdot\|_{D_m}^{-k})^\wedge(\xi) \left(\int_0^\infty t^{k-1}\phi(t\xi)\, dt\right) d\xi$$

$$= c\int_{S^{n-1}} \|\xi\|_L^{-n+k} \left(\int_0^\infty t^{k-1}\phi(t\xi)\, dt\right) d\xi$$

$$= c\int_{\mathbb{R}^n} \|y\|_L^{-n+k}\phi(y)\, dy.$$

Since ϕ is an arbitrary even test function, we get

$$(\|\cdot\|_D^{-k})^\wedge = c\|\cdot\|_L^{-n+k}.$$

□

Suppose that D is the k-intersection body of a star body. By Theorem 4.6, $(\|\cdot\|_D^{-k})^\wedge$ is a positive continuous, homogeneous of degree $-n+k$ function on $\mathbb{R}^n \setminus \{0\}$. For every even test function $\phi \in \mathcal{S}(\mathbb{R}^n)$, we have

$$\int_{\mathbb{R}^n} \|x\|_D^{-k}\phi(x)\, dx = \frac{1}{(2\pi)^n}\int_{\mathbb{R}^n} (\|\cdot\|_K^{-k})^\wedge(y)\hat{\phi}(y)\, dy$$

$$= \frac{1}{(2\pi)^n}\int_{S^{n-1}} (\|\cdot\|_D^{-k})^\wedge(\xi) \left(\int_0^\infty t^{k-1}\hat{\phi}(t\xi)\, dt\right) d\xi.$$

Similarly to how it was done in the case $k=1$ for intersection bodies, we define a more general class of bodies by replacing the positive continuous function $(\|\cdot\|_K^{-k})^\wedge$ in the latter equality by any measure on the sphere S^{n-1}.

Definition 4.7. *Let $1 \le k < n$. We say that an origin-symmetric star body D in $\mathbb{R}^n$ is a k-intersection body if there exists a finite Borel measure μ on S^{n-1} so that for every even test function ϕ in $\mathbb{R}^n$,*

$$\int_{\mathbb{R}^n} \|x\|_D^{-k}\phi(x)\, dx = \int_{S^{n-1}} \left(\int_0^\infty t^{k-1}\hat{\phi}(t\xi)\, dt\right) d\mu(\xi). \tag{4.12}$$

By the calculation before the definition, every k-intersection body of a star body is also a k-intersection body in the general sense.

We now present a Fourier analytic characterization of k-intersection bodies similar to Theorem 4.1.

Theorem 4.8. *Let $1 \le k < n$. An origin-symmetric star body D in $\mathbb{R}^n$ is a k-intersection body if and only if the function $\|\cdot\|_D^{-k}$ represents a positive definite distribution on $\mathbb{R}^n$.*

Proof. Suppose that D is a k-intersection body. Consider a non-negative even test function ϕ. Then, by (4.12) and the fact that $(\hat{\phi})^\wedge = (2\pi)^n\phi$, we get

$$\langle (\|\cdot\|_D^{-k})^\wedge, \phi\rangle = \langle \|\cdot\|_D^{-k}, \hat{\phi}\rangle = \int_{\mathbb{R}^n} \|x\|_D^{-k}\hat{\phi}(x)\, dx$$

$$= (2\pi)^n \int_{S^{n-1}} \left(\int_0^\infty t^{k-1}\phi(t\xi)dt \right) d\mu(\xi) \geq 0,$$

so the function $\|\cdot\|_D^{-k}$ represents a positive definite distribution.

On the other hand, if $\|\cdot\|_D^{-k}$ is a positive definite distribution, by Corollary 2.26, there exists a measure μ_0 on the sphere S^{n-1} so that for any even test function ϕ,

$$\int_{\mathbb{R}^n} \|x\|_D^{-k}\phi(x)\, dx = \int_{S^{n-1}} \left(\int_0^\infty t^{k-1}\hat{\phi}(t\xi)dt \right) d\mu_0(\xi),$$

which means that D is a k-intersection body. □

We are going to prove now that every origin-symmetric convex body in $\mathbb{R}^n$ is a k-intersection body for each $k = n-3, n-2, n-1$ such that $k > 0$. The result of Corollary 4.9 is actually more general, so we start with a simple argument in the infinitely smooth case, which is intended to show what is really going on.

Suppose that D is an infinitely smooth origin-symmetric convex body in $\mathbb{R}^n$. Put $k = 2$, $n \neq 3$ in the first formula of Theorem 3.18. We get that for every $\xi \in S^{n-1}$,

$$(\|\cdot\|_D^{-n+3})^\wedge(\xi) = -\pi(n-3)A''_{D,\xi}(0).$$

Then use the second formula of Theorem 3.18 with $k = 1$, $n \neq 2$:

$$(\|\cdot\|_D^{-n+2})^\wedge(\xi) = -2(n-2)\int_0^\infty \frac{A_{D,\xi}(z) - A_{D,\xi}(0)}{z^2}dz.$$

Finally, the first formula of Theorem 3.18 with $k = 0$ gives

$$(\|\cdot\|_D^{-n+1})^\wedge(\xi) = \pi(n-1)A_{D,\xi}(0) \geq 0.$$

Since the body D is symmetric convex and smooth, by Brunn's theorem, Theorem 2.3, for every $\xi \in S^{n-1}$, $A''_{D,\xi}(0) \leq 0$, and also $A_{D,\xi}(z) \leq A_{D,\xi}(0)$ for every $z \in \mathbb{R}$. We conclude that homogeneous locally integrable functions $\|\cdot\|_D^{-n+3}$, $\|\cdot\|_D^{-n+2}$ and $\|\cdot\|_D^{-n+1}$ represent positive definite distributions, provided that $n \geq 4$, $n \geq 3$, $n \geq 2$, respectively. By Theorem 4.8, D is a k-intersection body for each $k = n-3, n-2, n-1$ such that $k > 0$.

We now prove the result in its full strength.

COROLLARY 4.9. *For every origin-symmetric convex body K in $\mathbb{R}^n$ and every $n-3 \leq p < n$, $p > 0$, the function $\|\cdot\|_K^{-p}$ represents a positive definite distribution. In particular, every such body K is a k-intersection body for each $k = n-3, n-2, n-1$ such that $k > 0$.*

PROOF. By (2.43), for $\beta \in (-1, 0)$, the fractional derivative of the order β of the parallel section function $A_{K,\xi}$ at zero is equal to

$$A_{K,\xi}^{(\beta)}(0) = \frac{1}{\Gamma(-\beta)} \int_0^\infty t^{-1-\beta} A_{K,\xi}(t)\, dt. \tag{4.13}$$

If $\beta \in (0, 2)$, $\beta \neq 1$, then, by (2.42),

$$A_{K,\xi}^{(\beta)}(0) = \frac{1}{\Gamma(-\beta)} \int_0^\infty t^{-1-\beta}(A_{K,\xi}(t) - A_{K,\xi}(0))\, dt. \tag{4.14}$$

First assume that K is infinitely smooth and p is not an integer. Put $\beta = n - p - 1 \in (-1, 2)$. We are going to show that $(\|\cdot\|_K^{-n+\beta+1})^\wedge$ is a non-negative continuous function on S^{n-1}. Since this function is also homogeneous of degree $-\beta - 1 > -n$ on $\mathbb{R}^n$, we deduce that it is non-negative and locally integrable on

$\mathbb{R}^n$ (by Lemma 2.1). This means, in particular, that $(\|\cdot\|_K^{-n+\beta+1})^\wedge = (\|\cdot\|_K^{-p})^\wedge$ is a positive distribution, and $\|\cdot\|_K^{-p}$ is positive definite.

By Lemma 3.16, the restriction of $(\|\cdot\|_K^{-n+\beta+1})^\wedge$ to the sphere belongs to $C^\infty(S^{n-1})$. Let us show that $(\|\cdot\|_K^{-n+\beta+1})^\wedge$ is a non-negative function. To do this, we use (3.13) (recall that, by our assumption, β is not an integer):

$$A_{K,\xi}^{(\beta)}(0) = \frac{\cos(\pi\beta/2)}{\pi(n-\beta-1)}(\|\cdot\|_K^{-n+\beta+1})^\wedge(\xi). \tag{4.15}$$

First let $p \in (n-1, n)$. Then $\beta \in (-1, 0)$, so $\Gamma(-\beta) > 0$ and, by (4.13), $A_{K,\xi}^{(\beta)}(0) > 0$ for every $\xi \in S^{n-1}$. Also $\cos\frac{\beta\pi}{2} > 0$, so (4.15) implies non-negativity of the Fourier transform.

If $p \in (n-2, n-1)$, then $\beta \in (0,1)$, so $\Gamma(-\beta) < 0$. Since K is a convex body, by Brunn's theorem, the function $A_{K,\xi}$ has maximum at zero. Therefore, the integral in (4.14) is less than or equal to zero, and again $A_{K,\xi}^{(\beta)}(0) \geq 0$ for every $\xi \in S^{n-1}$. Also $\cos\frac{\beta\pi}{2} > 0$, so the result follows from (4.15).

If $p \in (n-3, n-2)$, then $\beta \in (1,2)$, so $\Gamma(-\beta) > 0$. The integral in (4.14) is less than or equal to zero for the same reason as in the case $\beta \in (0,1)$, so $A_{K,\xi}^{(\beta)}(0) \leq 0$ for every $\xi \in S^{n-1}$. But now $\cos\frac{\beta\pi}{2} < 0$.

Now we have to free ourselves from the restrictions on K imposed earlier in the proof. Suppose that p is still not an integer, but now K is not infinitely smooth. We can approximate K in the radial metric by infinitely smooth convex bodies K_m, $m \in \mathbb{N}$ (see Theorem 2.10). Let $\|\cdot\|_m$ be the norm on $\mathbb{R}^n$ with the unit ball K_m. By the first part of Lemma 3.11, for every non-negative test function ϕ and every $p \in [n-3, n)$,

$$\langle(\|\cdot\|_K^{-p})^\wedge, \phi\rangle = \lim_{m\to\infty}\langle(\|\cdot\|_m^{-p})^\wedge, \phi\rangle \geq 0,$$

because we have already proved that the distributions $\|\cdot\|_m^{-p}$ are positive definite.

Finally, the second part of Lemma 3.11 allows us to remove the assumption that p is not an integer.

The last statement of the corollary follows from Theorem 4.8. □

We end this section by proving a simple approximation argument that will later be used in the solution to the Busemann-Petty problem.

LEMMA 4.10. *Let $1 \leq k < n$. Suppose that D is an origin-symmetric convex body in $\mathbb{R}^n$ that is not a k-intersection body. Then there exists a sequence D_m of origin-symmetric convex bodies so that D_m converges to D in the radial metric, each D_m is infinitely smooth, has strictly positive curvature and each D_m is not a k-intersection body.*

PROOF. Consider a sequence of origin-symmetric infinitely smooth convex bodies D_m with positive curvature that approximate D in the radial metric (see the remark at the end of Section 2.2). Suppose that all the bodies D_m, starting from some m_0, are k-intersection bodies. Then $\|\cdot\|_{D_m}^{-k}$ is a positive definite distribution for each $m > m_0$. By Lemma 3.11, for every even test function ϕ,

$$\langle(\|\cdot\|_D^{-k})^\wedge, \phi\rangle = \lim_{m\to\infty}\langle(\|\cdot\|_{D_m}^{-k})^\wedge, \phi\rangle \geq 0,$$

which implies that D is a k-intersection body by Theorem 4.8. This gives a contradiction that proves our statement. □

4.3. L_p-balls as k-intersection bodies

The Fourier criterion for k-intersection bodies works well for unit balls of subspaces of L_p-spaces. First, we show that there is a large class of bodies that are k-intersection bodies for all k.

THEOREM 4.11. *The unit ball of any n-dimensional subspace of L_q with $0 < q \le 2$ is a k-intersection body for every $1 \le k < n$.*

PROOF. By a well-known result of P. Levy (see Theorem 6.6) for every n-dimensional subspace $B = (\mathbb{R}^n, \|\cdot\|)$ of L_q with $0 < q \le 2$, the function $\exp(-\|\cdot\|^q)$ is positive definite on $\mathbb{R}^n$. By Bochner's theorem, Theorem 6.2, this function is the Fourier transform of a finite Borel measure μ on $\mathbb{R}^n$.

Using the change of variables $z = t\|x\|$ and formula (2.26), for every $x \in \mathbb{R}^n$ we get

$$\|x\|^{-k} = \frac{q}{\Gamma(k/q)} \int_0^\infty t^{k-1} \exp(-t^q\|x\|^q)\, dt.$$

By Lemma 2.1, for any even test function ϕ, the function $\|\cdot\|^{-k}\hat{\phi}(\cdot)$ is integrable on $\mathbb{R}^n$. Therefore, we can use the Fubini theorem in the following calculation:

$$\langle (\|\cdot\|^{-k})^\wedge, \phi\rangle = \int_{\mathbb{R}^n} \|x\|^{-k}\hat{\phi}(x)\, dx$$

$$= \frac{q}{\Gamma(k/q)} \int_{\mathbb{R}^n} \hat{\phi}(x) \left(\int_0^\infty t^{k-1} \exp(-t^q\|x\|^q)\, dt \right) dx$$

$$= \frac{q}{\Gamma(k/q)} \int_0^\infty t^{k-1} \left(\int_{\mathbb{R}^n} \hat{\phi}(x) \exp(-t^q\|x\|^q)\, dx \right) dt$$

$$= \frac{q}{\Gamma(k/q)} \int_0^\infty t^{k-1} \left(\int_{\mathbb{R}^n} t^{-n}\hat{\phi}(y/t) \exp(-\|y\|^q)\, dy \right) dt. \tag{4.16}$$

The function $y \mapsto t^{-n}\hat{\phi}(y/t)$ is the Fourier transform of the function $x \mapsto \phi(tx)$. Therefore, since $\hat{\mu} = \exp(-\|\cdot\|^q)$ and $\langle \hat{f}, \hat{\phi}\rangle = (2\pi)^n \langle f, \phi\rangle$, we have

$$\int_{\mathbb{R}^n} t^{-n}\hat{\phi}(y/t) \exp(-\|y\|^q)\, dy = \langle \exp(-\|y\|^q), t^{-n}\hat{\phi}(y/t)\rangle$$

$$= (2\pi)^n \langle \mu, \phi(tx)\rangle = (2\pi)^n \int_{\mathbb{R}^n} \phi(tx)\, d\mu(x).$$

Now (4.16) can be written as

$$\langle (\|\cdot\|^{-k})^\wedge, \phi\rangle = \frac{(2\pi)^n q}{\Gamma(k/q)} \int_0^\infty t^{k-1} \left(\int_{\mathbb{R}^n} \phi(tx)\, d\mu(x) \right) dt.$$

The latter equality shows that $\langle (\|\cdot\|^{-k})^\wedge, \phi\rangle \ge 0$ if ϕ is any even non-negative test function, so $(\|\cdot\|^{-k})^\wedge$ is a positive distribution. If K is the unit ball of the space B, then $\|\cdot\|$ is the Minkowski functional of K, so K is a k-intersection body by Theorem 4.8. □

The situation is completely different if $q > 2$. We shall prove now that the unit balls of the spaces ℓ_q^n, $q > 2$, represent the extreme case – they are k-intersection bodies only for $k = n-3, n-2, n-1$ such that $k > 0$. Recall that, by Corollary 4.9, every origin-symmetric convex body in $\mathbb{R}^n$ is a k-intersection body for these values of k.

We start with the case $2 < q < \infty$.

LEMMA 4.12. *If* $q > 2,\ n > 3,\ p \in (0, n-3)$, *then the distribution* $\|\cdot\|_q^{-p}$ *is not positive definite.*

PROOF. Recall that, by Lemma 3.6, the Fourier transform of $\|\cdot\|_q^{-p}$ is equal to a locally integrable function

$$(\|\cdot\|_q^{-p})^\wedge(\xi) = \frac{q}{\Gamma(p/q)} \int_0^\infty t^{n-p-1} \prod_{k=1}^n \gamma_q(t\xi_k)\, dt. \tag{4.17}$$

We use the properties of the function γ_q; see Section 2.8. As before, for $-1 < z < q$, we denote

$$s_q(z) = \int_{\mathbb{R}} |t|^z \gamma_q(t)\ dt.$$

By Lemma 2.29, the moments $s_q(z)$ are positive for $z \in (-1,0)\cup(0,2)$ and negative for $z \in (2, \min(q,4))$.

Use formula (4.17) to compute the integral

$$I(z_1,\dots,z_{n-1}) = \int_{\mathbb{R}^{n-1}} |\xi_1|^{z_1} \dots |\xi_{n-1}|^{z_{n-1}} (\|\cdot\|_q^{-p})^\wedge(\xi_1, ..., \xi_{n-1}, 1)\ d\xi$$
$$= (q/\Gamma(p/q))\ s_q(z_1)\dots s_q(z_{n-1}) s_q(-z_1 - \dots - z_{n-1} - p).$$

By Lemma 2.28, this integral converges absolutely if each of the numbers

$$z_1, \dots, z_{n-1},\ -z_1 - \dots - z_{n-1} - p$$

belongs to the interval $(-1, q)$. Choosing $z_k \in (-1,0)$ for every $k = 1, ..., n-1$, we have the moments $s_q(z_k)$, $k = 1, ..., n-1$ positive, and we can make the number $-z_1 - \dots - z_{n-1} - p$ equal to any number from the interval $(-p, n-1-p) \cap (-1, q)$. Since $p < n-3$ and $q > 2$, this interval contains a neighborhood of 2. The moment function s_q changes its sign at 2, so we can make the integral $I(z_1, \dots, z_{n-1})$ positive for one choice of z's and negative for another choice. This means that the function $(\|\cdot\|_q^{-p})^\wedge$ is sign-changing. □

An immediate consequence of the previous lemma, Corollary 4.9 and Theorem 4.8 is the following theorem.

THEOREM 4.13. *Let* $0 < p < n,\ n \geq 3$. *If* $2 < q < \infty$, *then* $\|\cdot\|_q^{-p}$ *is a positive definite distribution if* $p \in [n-3, n)$, *and it is not positive definite if* $p \in (0, n-3)$. *Therefore the unit ball of the space* ℓ_q^n *is a* k*-intersection body if and only if* $n - 3 \leq k < n,\ k \neq 0$.

We now prove a similar result for $q = \infty$.

LEMMA 4.14. *Let* $q > 2,\ p \in (0,n)$. *Then, for every* $\xi \in \mathbb{R}^n$ *with non-zero coordinates,*

$$(\|\cdot\|_q^{-p})^\wedge(\xi)$$
$$= \frac{q^{n+1}}{p\Gamma(p/q)} \int_0^\infty \dots \int_0^\infty (t_1 \dots t_n)^q \exp(-\|t\|_q^q)\ (\|\cdot\|_\infty^{-p})^\wedge(t_1\xi_1, ..., t_n\xi_n)\ dt_1...dt_n.$$

PROOF. For every $x \in \mathbb{R}$, we have

$$\exp(-|x|^q) = q \int_0^\infty \chi(ux)\ u^{-1-q} \exp(-u^{-q})\ du, \tag{4.18}$$

where, as before, χ is the indicator of $[-1, 1]$.

If $u \in \mathbb{R}$, $u \neq 0$, then the Fourier transform of the function $x \mapsto \chi(ux)$ is equal to $(\chi(ux))^\wedge(\xi) = 2\sin(\xi/u)/\xi$. Calculating the Fourier transforms of both sides of (4.18), as functions of $x \in \mathbb{R}$, and making the change of variables $t = 1/u$, we get an integral representation for the function γ_q : for every $\xi \in \mathbb{R}$,

$$\gamma_q(\xi) = 2q \int_0^\infty \frac{\sin(t\xi)}{t\xi} t^q \exp(-t^q)\, dt.$$

Substituting this in (4.17), we get

$$(\|\cdot\|_q^{-p})^\wedge(\xi) = \frac{q}{\Gamma(p/q)} \int_0^\infty z^{n-p-1} \prod_{k=1}^n \gamma_q(z\xi_k)\, dz = \frac{2^n q^{n+1}}{\Gamma(p/q)}$$

$$\times \int_0^\infty \cdots \int_0^\infty (t_1 \dots t_n)^q \exp(-\|t\|_q^q) \left(\int_0^\infty z^{n-p-1} \prod_{k=1}^n \frac{\sin(t_k\xi_k z)}{t_k\xi_k z}\, dz \right) dt,$$

and the result follows from the formula for the Fourier transform of $\|\cdot\|_\infty^{-p}$ from Lemma 3.4. □

Since we already know that, for $2 < q < \infty$, $\|\cdot\|_q^{-p}$ is not positive definite if $0 < p < n-3$, we get the same result for $q = \infty$, because otherwise Lemma 4.14 leads to a contradiction. In conjunction with Theorem 4.8 and Corollary 4.9, this proves that the cube in $\mathbb{R}^n$ is a k-intersection body if and only if $k = n-3, n-2, n-1$ with $k > 0$. In particular, the four-dimensional cube is an intersection body, but the five-dimensional cube is not.

THEOREM 4.15. *Let $0 < p < n$, $n \geq 3$. The function $\|\cdot\|_\infty^{-p}$ is a positive definite distribution if $p \in [n-3, n)$, and it is not positive definite if $p \in (0, n-3)$. In particular, the unit cube in $\mathbb{R}^n$ is a k-intersection body in $\mathbb{R}^n$ if and only if $n-3 \leq k < n$, $k \neq 0$.*

We now give a different proof of the latter theorem. It is possible to compute the Fourier transform of any function of the ℓ_∞-norm, which gives a different expression for the Fourier transform of $\|\cdot\|_\infty^{-p}$. Besides, we are going to use the calculation below in the solution of the second Schoenberg problem in Section 6.2.

Denote by G the set of vectors $\xi = (\xi_1, \dots, \xi_n) \in \mathbb{R}^n$ such that $\xi_k \neq 0$ for $1 \leq k \leq n$ and the scalar product $(\delta, \xi) \neq 0$ for every vector $\delta = (\delta_1, \dots, \delta_n)$, with $\delta_k = \pm 1$, $1 \leq k \leq n$.

LEMMA 4.16. *Let f be an even continuous function on $\mathbb{R}$ with power growth at infinity (i.e. there exist $A, \beta > 0$ so that $|f(t)| < A(1+|t|)^\beta$ for every $t \in \mathbb{R}$) so that the distribution $u = (f(t)(\mathrm{sgn}(t))^{n-1})^\wedge$ is a continuous function on $\mathbb{R} \setminus \{0\}$. Then, the Fourier transform of $f(\|\cdot\|_\infty)$ is equal in G to a continuous function*

$$\xi \mapsto \frac{i^{n-1}}{2\xi_1 \cdot \ldots \cdot \xi_n} \sum_\delta \delta_1 \cdot \ldots \cdot \delta_n\ (\delta, \xi)\ u((\delta, \xi)), \qquad \xi \in G,$$

where the sum is taken over all changes of signs $\delta = (\delta_1, \dots, \delta_n)$, $\delta_j = \pm 1$, $j = 1, \dots, n$.

PROOF. Let ϕ be a test function with compact support in G. Then there exists a test function F so that $\partial^n F/\partial x_1 \dots \partial x_n = \hat{\phi}$. In fact, by the definition of the set G the function ϕ is supported outside of the coordinate planes, so the function

$\psi(x) = \phi(x)/(x_1 \dots x_n)$ belongs to $\mathcal{S}(\mathbb{R}^n)$, and hence $\hat{\psi} \in \mathcal{S}(\mathbb{R}^n)$. Put $F = i^n \hat{\psi}$ and use the connection between the Fourier transform and differentiation, (2.36):

$$\partial^n F/\partial x_1 \dots \partial x_n = i^n \partial^n \hat{\psi}/\partial x_1 \dots \partial x_n = (x_1 \dots x_n \psi)^\wedge = \hat{\phi}.$$

If ϕ is an even function, then the function F is even if n is an even integer, and it is an odd function if n is an odd integer.

Let $B_t = \{x \in \mathbb{R}^n : \|x\|_\infty < t\}$. Consider a set

$$Q_h = \{(x,t) \in \mathbb{R}^n \times \mathbb{R} : \ t \le \|x\|_\infty \le t+h\}.$$

Applying the Fubini theorem to the integral

$$\frac{1}{h} \int_{Q_h} f(t)\phi(x)\ dx\ dt$$

and sending h to zero, we get the well-known formula

$$\int_{\mathbb{R}^n} f(\|x\|_\infty)\phi(x)\ dx = \int_{\mathbb{R}} f(t) \left(\int_{B_t} \phi(x)\ dx \right)'_t\ dt$$

that we use in the calculation below.

Now we start the calculation: for every even test function ϕ with compact support in G,

$$\begin{aligned}
\langle (f(\|\cdot\|_\infty))^\wedge, \phi \rangle &= \langle f(\|\cdot\|_\infty), \hat{\phi} \rangle \\
&= \int_{\mathbb{R}^n} f(\|x\|_\infty) \hat{\phi}(x)\, dx \\
&= \int_0^\infty f(t) \left(\int_{B_t} \hat{\phi}(x)\, dx \right)'_t dt \\
&= \int_0^\infty f(t) \left(\int_{-t}^t \dots \int_{-t}^t \hat{\phi}(x)\, dx \right)'_t dt \\
&= \int_0^\infty f(t) \sum_\delta \delta_1 \cdot \ldots \cdot \delta_n (F(\delta_1 t, \dots, \delta_n t))'_t\, dt \\
&= \int_0^\infty f(t) \left(\sum_\delta \delta_1 \cdot \ldots \cdot \delta_n \left(\delta_1 \frac{\partial F}{\partial x_1} + \dots + \delta_n \frac{\partial F}{\partial x_n} \right) (\delta_1 t, \dots, \delta_n t) \right) dt.
\end{aligned}$$

The function in parentheses is even if n is an odd integer and odd if n is even. Therefore, the latter integral can be written as the action of a distribution on a test function as follows:

$$\text{(4.19)} \quad \frac{1}{2} \Big\langle f(t)(\operatorname{sgn}(t))^{n-1}, \sum_\delta \delta_1 \cdot \ldots \cdot \delta_n \left(\delta_1 \frac{\partial F}{\partial x_1} + \dots + \delta_n \frac{\partial F}{\partial x_n} \right) (\delta_1 t, ..., \delta_n t) \Big\rangle.$$

Since $\partial F/\partial x_j = i^{n-1}(x_j \phi(x)/(x_1 \dots x_n))^\wedge$ for each j, we can use the connection between the Radon transform and the Fourier transform, Lemma 2.11, to write (4.19) in the form

$$\begin{aligned}
&\frac{i^{n-1}}{2} \Big\langle f(t)(\operatorname{sgn}(t))^{n-1}, \sum_\delta \delta_1 \cdot \ldots \cdot \delta_n \left(\sum_{j=1}^n \int_{(\delta,x)=y} \frac{\delta_j x_j \phi(x)}{x_1 \dots x_n}\, dx \right)^\wedge (t) \Big\rangle \\
&= \frac{i^{n-1}}{2} \Big\langle (f(t)(\operatorname{sgn}(t))^{n-1})^\wedge(y), y \sum_\delta \delta_1 \cdot \ldots \cdot \delta_n \int_{(\delta,x)=y} \frac{\phi(x)}{x_1 \dots x_n}\, dx \Big\rangle.
\end{aligned}$$

The distribution $u = (f(t)(\mathrm{sgn}(t))^{n-1})^\wedge$ is a continuous function on $\mathbb{R} \setminus \{0\}$; therefore for every $\phi \in \mathcal{S}(\mathbb{R}^n)$ with compact support in G we have

$$\text{(4.20)}\quad \langle f(\|\cdot\|_\infty)^\wedge, \phi\rangle = \frac{i^{n-1}}{2}\int_{\mathbb{R}}\left(\sum_\delta \delta_1\cdot\ldots\cdot\delta_n \int_{(\delta,x)=y}\frac{\phi(x)}{x_1\ldots x_n}\,dx\right) y\; u(y)\,dy.$$

The latter integral converges absolutely because all the functions

$$y \to \int_{(\delta,x)=y}\frac{\phi(x)}{x_1\ldots x_n}\,dx, \qquad y \in \mathbb{R},$$

belong to $\mathcal{S}(\mathbb{R})$ and have compact support in $\mathbb{R}\setminus\{0\}$. By the Fubini theorem, (4.20) is equal to

$$\frac{i^{n-1}}{2}\int_{\mathbb{R}^n}\left(\sum_\delta \delta_1\cdot\ldots\cdot\delta_n\frac{(\delta,x)}{x_1\ldots x_n}u((\delta,x))\right)\phi(x)\,dx,$$

which completes the proof. □

The Fourier transform of the function $t \mapsto |t|^p,\ t \in \mathbb{R}$, was computed in Lemma 2.23 (with the coefficient in the form of (2.40)): if p is not an integer, then, for every $y \in \mathbb{R}\setminus\{0\}$

$$(|t|^p)^\wedge(y) = \frac{2^{p+1}\sqrt{\pi}\,\Gamma((p+1)/2)}{\Gamma(-p/2)}|y|^{-1-p}.$$

Let us use this formula to compute the Fourier transform of the function $|t|^p\mathrm{sgn}(t)$. This function is the derivative of $|t|^{p+1}$, so, by the connection between the Fourier transform and differentiation, we get

$$(|t|^p\mathrm{sgn}(t))^\wedge(y) = \frac{1}{p+1}\left((|t|^{p+1})'\right)^\wedge(y)$$

$$= \frac{i}{p+1}\,y\,(|t|^{p+1})^\wedge(y).$$

By the formula for the Fourier transform of $|t|^p$, the latter is equal to

$$\frac{i2^{p+2}\sqrt{\pi}\Gamma((p+2)/2)}{(p+1)\Gamma((-p-1)/2)}\,y\,|y|^{-p-2}.$$

Finally, using the fact that $\Gamma(x+1) = x\Gamma(x)$ (see (2.27)), we get

$$(|t|^p\mathrm{sgn}(t))^\wedge(y) = \frac{-i2^{p+1}\sqrt{\pi}\,\Gamma((p+2)/2)}{\Gamma((-p+1)/2)}|y|^{-1-p}\mathrm{sgn}(y).$$

Let us apply the formula of Lemma 4.16 to the functions $f(t) = |t|^p$, where $p > 0$, p is not an integer. Computing the function u according to the formulas above, we get new formulas for the Fourier transform of powers of the ℓ_∞-norm. By Lemma 2.22, these formulas can be extended analytically by p to negative values of $p > -n$ that are not integers. We get the following lemma.

LEMMA 4.17. *Suppose $0 < p < n$, p is not an integer, and $\xi \in \mathbb{R}^n$ is a vector from G. Then, if n is odd, we have*

$$(\|\cdot\|_\infty^{-p})^\wedge(\xi) = \frac{(-1)^{\frac{n-1}{2}}2^{-p}\sqrt{\pi}\,\Gamma(\frac{-p+1}{2})}{\xi_1\ldots\xi_n\,\Gamma(p/2)}\sum_\delta \delta_1\ldots\delta_n\,|(\delta,\xi)|^p\mathrm{sgn}((\delta,\xi)).$$

If n is even,

$$(\|\cdot\|_\infty^{-p})^\wedge(\xi) = \frac{(-1)^{\frac{n}{2}+1}2^{-p}\sqrt{\pi}\ \Gamma((-p+2)/2)}{\xi_1\dots\xi_n\ \Gamma((p+1)/2)}\sum_\delta \delta_1\dots\delta_n\ |(\delta,\xi)|^p,$$

where the sum is taken over all choices of signs $\delta = (\delta_1,\dots,\delta_n)$, $\delta_k = \pm1$, $k = 1,...,n$.

It is possible to get the result of Lemma 4.17 by performing integration in the formula of Lemma 3.4. First let $-1 < p < 0$ and use the representation of the product $\prod_{k=1}^n \sin(t\xi_k)$ as a sign-changing sum of sines or cosines. Then extend the formula analytically by p. This allows us to double-check the result of Lemma 4.17.

For $n \geq 2$ and $p \in \mathbb{R}$ let

$$u_{n,p}(\xi_1,...,\xi_n) = \sum_\delta \delta_1\dots\delta_n\ |(\delta,\xi)|^p\mathrm{sgn}((\delta,\xi))$$

if n is an odd integer, and

$$u_{n,p}(\xi_1,...,\xi_n) = \sum_\delta \delta_1\dots\delta_n\ |(\delta,\xi)|^p$$

if n is an even integer.

LEMMA 4.18. *Let $n > 3$ and $0 < p < n-3$, or $n = 3$ and $p < 0$, where p is not an integer. Then the function $u_{n,p}$ is sign-changing on $\mathbb{R}^n_+ = \{\xi \in \mathbb{R}^n : \xi_k > 0,\ k = 1,...,n\}$.*

PROOF. First, let $n = 3$. Then, for every $p < 0$, the number

$$u_{3,p}(3,1,1) = 5^p - 2\cdot 3^p + 1^p$$

is positive (the function $|t|^p$ is convex on $(0,\infty)$), but the number $u_{3,p}(1,3,3)$ is negative.

Now we deduce the result for $n > 3$ and $0 < p < n-3$ from the three-dimensional case. In fact, put $\xi_{n-2} = 3$, $\xi_{n-1} = \xi_n = 1$. Then the limit

$$\lim_{\xi_1\to+0,\dots,\xi_{n-3}\to+0} \frac{1}{\xi_1\dots\xi_{n-3}} u_{n,p}(\xi)$$

is equal, up to a non-zero constant, to $u_{3,p-n+3}(3,1,1)$. If $\xi_{n-2} = 1$, $\xi_{n-1} = \xi_n = 3$, the same limit is equal (up to the same non-zero constant) to $u_{3,p-n+3}(1,3,3)$. Since $p-n+3 < 0$, this shows that, if we make $\xi_1,\dots,\xi_{n-3}$ small enough, the values of the function $u_{n,p}$ at the points $(\xi_1,\dots,\xi_{n-3},3,1,1)$ and $(\xi_1,\dots,\xi_{n-3},1,3,3)$ have opposite signs. □

The results of Lemmas 4.17 and 4.18 confirm the claim of Theorem 4.15 for $0 < p < n-3$, where p is not an integer.

4.4. The second derivative test

The fact that the unit balls of ℓ_q-spaces with $q > 2$ are not k-intersection bodies for $k < n-3$ is a part of a more general result that we prove in this section. We are going to show that the same is true for essentially any norm, whose second directional derivative at zero vanishes in a certain way. These norms include q-sums of Banach spaces, where $q > 2$, and Orlicz norms with the Orlicz function M such that $M''(0) = 0$.

The idea behind the following result is quite simple, though its realization requires a substantial technical effort. Suppose that f is a positive definite function on $\mathbb{R}$. Then, by Bochner's theorem, Theorem 6.2, f is the Fourier transform of a probability Borel measure μ on $\mathbb{R}$; see Section 6.1. Under rather general conditions on the function f, we can use the connection between the Fourier transform and differentiation:

$$f''(0) = -\int_{\mathbb{R}} t^2\, d\mu(t).$$

Now if the second derivative of f vanishes at zero, we get that the measure μ is supported in the origin, so $f \equiv 1$. This argument is hidden in the proof of the following result:

THEOREM 4.19. *Let* $0 < p < n$ *and let* X *be an* n*-dimensional normed space with a normalized basis* $e_1, \dots, e_n$, $n \geq 3$, *so that:*
(i) For every fixed $(x_2, \dots, x_n) \in \mathbb{R}^{n-1} \setminus \{0\}$, *the function*

$$x_1 \mapsto \|x_1 e_1 + \sum_{i=2}^{n} x_i e_i\|$$

has a continuous second derivative everywhere on $\mathbb{R}$, *and*

$$\|x\|'_{x_1}(0, x_2, \dots, x_n) = \|x\|''_{x_1^2}(0, x_2, \dots, x_n) = 0,$$

where $\|x\|'_{x_1}$ *and* $\|x\|''_{x_1^2}$ *stand for the first and second partial derivatives by* x_1 *of the norm* $\|x_1 e_1 + \dots + x_n e_n\|$.
(ii) There exists a constant C *so that, for every* $x_1 \in \mathbb{R}$ *and every* $(x_2, \dots, x_n) \in \mathbb{R}^{n-1}$ *with* $\|x_2 e_2 + \dots + x_n e_n\| = 1$, *one has*

$$\|x\|''_{x_1^2}(x_1, x_2, \dots, x_n) \leq C.$$

(iii) Convergence in the limit

$$\lim_{x_1 \to 0} \|x\|''_{x_1^2}(x_1, x_2, \dots, x_n) = 0$$

is uniform with respect to $(x_2, \dots, x_n) \in \mathbb{R}^{n-1}$ *with* $\|x_2 e_2 + \dots + x_n e_n\| = 1$.

Then the function $\|\cdot\|^{-p}$ *represents a positive definite distribution if and only if* $p \in [n-3, n)$. *In particular the convex set* $\{x \in \mathbb{R}^n : \|\sum_{i=1}^n x_i e_i\| \leq 1\}$ *is a* k*-intersection body if and only if* k *is one of the numbers* $n-3, n-2, n-1$, $k \neq 0$.

Throughout this section, for each $x = (x_1, ..., x_n) \in \mathbb{R}^n$ we denote by y the vector $(x_2, ..., x_n) \in \mathbb{R}^{n-1}$. We need several remarks:

(i) By Lemma 2.1, if $p < n-3$, the function $y \mapsto \|y\|^{-p-2}$ is locally integrable on $\mathbb{R}^{n-1}$. This is the crucial point in the proof of Lemma 4.20 below. Note that this is no longer true for $p \geq n-3$.

(ii) A simple consequence of the triangle inequality is that $-1 \leq \|x\|'_{x_1} \leq 1$ at every point $x \in \mathbb{R}^n$ with $y \neq 0$.

(iii) For every fixed $y \in \mathbb{R}^{n-1} \setminus \{0\}$, $x_1 \mapsto \|x_1 e_1 + \sum_{i=2}^n x_i e_i\|$ is a convex differentiable function of x_1 whose derivative at zero is equal to zero. Therefore, for every $x = (x_1, \dots, x_n) \in \mathbb{R}^n$, we have $\|x\| \geq \|y\|$.

(iv) The function $\|x\|''_{x_1^2}$ is non-negative and homogeneous of degree -1. Let C be the constant from condition (ii) of Theorem 4.19. Then, for every $x \in \mathbb{R}^n$ with $y \in \mathbb{R}^{n-1} \setminus \{0\}$, we have

$$\|x\|''_{x_1^2}(x) = (1/\|y\|)\|x\|''_{x_1^2}(x/\|y\|) \leq C/\|y\|.$$

We start the proof of Theorem 4.19. For every $m \in \mathbb{N}$, put

$$h_m(x_1) = (m/\sqrt{2\pi}) \exp(-x_1^2 m^2/2).$$

Clearly,

$$\int_{\mathbb{R}} h_m(x_1)\, dx_1 = 1,$$

and for every $\delta > 0$,

$$\lim_{m\to\infty} \int_{|x_1|>\delta} h_m(x_1)\, dx_1 = 0.$$

Now for $(x_2, ..., x_n) \in \mathbb{R}^{n-1}$, let

$$u(x_2, ..., x_n) = (1/(2\pi)^{(n-1)/2}) \exp(-(x_2^2 + ... + x_n^2)/2),$$

and consider test functions

$$\phi_m(x) = h_m(x_1) u(x_2, ..., x_n) \in \mathcal{S}(\mathbb{R}^n).$$

Since for large m these functions are essentially supported where x_1 is small, and the first and second derivatives of the norm by x_1 approach zero when $x_1 \to 0$, one can expect the quantity $\langle (\|x\|^{-p})''_{x_1^2}\ , \phi_m\rangle$ to approach zero as $m \to \infty$. We need a weaker fact:

LEMMA 4.20. *Let $0 < p < n - 3$. For every $\epsilon > 0$ there exists $N \in \mathbb{N}$ so that, for every $m > N$, $\langle (\|x\|^{-p})''_{x_1^2}\ , \phi_m\rangle \geq -\epsilon$.*

PROOF. First we use the definition of the derivative of a distribution, Fubini's theorem and the fact that the function $\|\cdot\|^{-p}$ is locally integrable to show that

$$\langle (\|x\|^{-p})''_{x_1^2}\ , \phi_m\rangle = \langle \|\cdot\|^{-p}, \frac{\partial^2 \phi_m}{\partial x_1^2}\rangle = \int_{\mathbb{R}^n} \|x\|^{-p} \frac{\partial^2 \phi_m}{\partial x_1^2}\, dx$$

$$= \int_{\mathbb{R}^{n-1}} u(y) \left(\int_{\mathbb{R}} \|x\|^{-p} \frac{\partial^2 h_m}{\partial x_1^2}\, dx_1 \right) dy$$

$$= \int_{\mathbb{R}^{n-1}} \langle (\|x\|^{-p})''_{x_1^2}(x)\ , h_m\rangle u(y)\, dy. \tag{4.21}$$

By conditions (i) and (ii) of Theorem 4.19, for every fixed $y \in \mathbb{R}^{n-1} \setminus \{0\}$, the derivative $(\|x\|^{-p})''_{x_1^2}$ is a continuous bounded function of the variable x_1 on $\mathbb{R}$, so the integral in (4.21) can be written as

$$\int_{\mathbb{R}^{n-1}} \left(\int_{\mathbb{R}} (\|x\|^{-p})''_{x_1^2}(x) h_m(x_1)\, dx_1 \right) u(y)\, dy$$

$$= \int_{\mathbb{R}^{n-1}} \left(\int_{\mathbb{R}} p(p+1)\|x\|^{-p-2} \left(\|x\|'_{x_1}(x)\right)^2 h_m(x_1)\, dx_1 \right) u(y)\, dy$$

$$- \int_{\mathbb{R}^{n-1}} \left(\int_{\mathbb{R}} p\|x\|^{-p-1} \|x\|''_{x_1^2}(x) h_m(x_1)\, dx_1 \right) u(y)\, dy.$$

The first integral is non-negative, so, to prove Lemma 4.20, it suffices to show that

$$\lim_{m\to\infty} \int_{\mathbb{R}^n} \|x\|^{-p-1} \|x\|''_{x_1^2}(x)\, h_m(x_1)\, u(y)\, dx = 0.$$

Let $\epsilon > 0$. By remark (i), the function $\|y\|^{-p-2}$ is locally integrable on $\mathbb{R}^{n-1}$, so, by Lemma 2.1,

$$L = \int_{\mathbb{R}^{n-1}} \|y\|^{-p-2} u(y)\, dy < \infty.$$

Besides, there exists $d > 0$ so that

$$\int_{\{y:\|y\|<d\}} \|y\|^{-p-2} u(y)\, dy < \frac{\epsilon}{3C}, \tag{4.22}$$

where C is the number from condition (ii) of Theorem 4.19. By remark (iv), we have

$$\int_{\mathbb{R}\times\{y:\|y\|<d\}} \|x\|^{-p-1} \|x\|''_{x_1^2}(x)\; h_m(x_1)\; u(y)\, dx$$

$$\leq C \int_{\mathbb{R}\times\{y:\|y\|<d\}} \|x\|^{-p-1} \|y\|^{-1} h_m(x_1) u(y)\, dx < \frac{\epsilon}{3}, \tag{4.23}$$

where we applied remark (iii) to get that $\|x\|^{-p-1} \leq \|y\|^{-p-1}$, and then we used inequality (4.22) and the fact that $\int_{\mathbb{R}} h_m(t)\, dt = 1$.

By condition (iii) of Theorem 4.19, $\lim_{x_1 \to 0} \|x\|''_{x_1^2}(x) = 0$ uniformly with respect to $y \in \mathbb{R}^{n-1}$ with $\|y\| = 1$, so there exists $\delta > 0$ so that if $|x_1| < \delta$, then $\|x\|''_{x_1^2}(x) < \frac{\epsilon}{3L}$ for every $y \in \mathbb{R}^{n-1}$ with $\|y\| = 1$. The second derivative of the norm is a homogeneous function of degree -1. Therefore, if $y \in \mathbb{R}^{n-1} \setminus \{0\}$ and $\frac{|x_1|}{\|y\|} < \delta$, then

$$\|x\|''_{x_1^2}(x) = \frac{1}{\|y\|} \|x\|''_{x_1^2}\left(\frac{x}{\|y\|}\right) < \frac{\epsilon}{3L\|y\|}. \tag{4.24}$$

Consider the sets $A_1 = \{x_1 : |x_1| < \delta d\} \times \{y : \|y\| \geq d\}$ and $A_2 = \{x_1 : |x_1| \geq \delta d\} \times \{y : \|y\| \geq d\}$ in $\mathbb{R}^n$. For every point in A_1 we have $|x_1|/\|y\| < \delta$. Therefore, by (4.24), remark (iii) and the fact that $\int_{\mathbb{R}} h_m(t)\, dt = 1$, we get

$$\int_{A_1} \|x\|^{-p-1} \|x\|''_{x_1^2}(x)\; h_m(x_1)\; u(y)\, dx$$

$$\leq \frac{\epsilon}{3L} \int_{A_1} \|y\|^{-p-2} h_m(x_1)\; u(y)\, dx \leq \frac{\epsilon}{3}. \tag{4.25}$$

Finally, suppose that m is large enough so that $\int_{|x_1|\geq \delta d} h_m(x_1)\, dx_1 < \epsilon/(3CL)$. Then we use the estimate of remark (iv) and remark (iii) to show that

$$\int_{A_2} \|x\|^{-p-1} \|x\|''_{x_1^2}(x)\; h_m(x_1)\; u(y)\, dx$$

$$\leq C \int_{A_2} \|y\|^{-p-2} h_m(x_1)\; u(y)\, dx \leq \frac{\epsilon}{3}. \tag{4.26}$$

Since ϵ is an arbitrary positive number, the result of Lemma 4.20 follows from (4.23), (4.25), (4.26). □

We now finish the proof of Theorem 4.19. As follows from Corollary 4.9, for every n-dimensional normed space and every $p \in [n-3, n)$, $p > 0$, the function $\|\cdot\|^{-p}$ represents a positive definite distribution on $\mathbb{R}^n$.

Let $0 < p < n-3$. Suppose that $\|\cdot\|^{-p}$ is positive definite. Then, by Corollary 2.26, there exists a finite Borel measure μ on the sphere S^{n-1} so that for every even test function ϕ

$$(4.27)\qquad \langle \|\cdot\|^{-p}, \phi\rangle = \int_{\mathbb{R}^n} \|x\|^{-p}\phi(x)\, dx = \int_{S^{n-1}} \Big(\int_{\mathbb{R}} |t|^{p-1}\hat{\phi}(t\xi)\, dt \Big)\, d\mu(\xi).$$

By the connection between the Fourier transform and differentiation, we have $(\partial^2\phi/\partial x_1^2)^\wedge(\xi) = -\xi_1^2\hat{\phi}(\xi)$, and it follows from (4.27) that

$$(4.28)\quad \langle (\|x\|^{-p})''_{x_1^2}, \phi\rangle = \langle \|\cdot\|^{-p}, \frac{\partial^2\phi}{\partial x_1^2}\rangle = -\int_{S^{n-1}} \Big(\int_{\mathbb{R}} |t|^{p+1}\hat{\phi}(t\xi)\, dt \Big)\ \xi_1^2\, d\mu(\xi).$$

Let ϕ_m be the function defined just before Lemma 4.20. Then for every $\xi \in \mathbb{R}^n$

$$\hat{\phi}_m(\xi) = \exp(-\frac{\xi_1^2}{2m^2}) \exp(-(\xi_2^2 + \cdots + \xi_n^2)/2).$$

Applying (4.28) to the functions ϕ_m and using (2.26), we get

$$\langle (\|x\|^{-p})''_{x_1^2}, \phi_m\rangle = -\frac{2^{p/2}}{\Gamma((p+2)/2)} \int_{S^{n-1}} \xi_1^2 (\frac{\xi_1^2}{m^2} + \xi_2^2 + \cdots + \xi_n^2)^{-\frac{p+2}{2}}\, d\mu(\xi)$$

$$\le -\frac{2^{p/2}}{\Gamma((p+2)/2)} \int_{S^{n-1}} \xi_1^2\, d\mu(\xi),$$

because $\xi \in S^{n-1}$ and $m \ge 1$.

By Lemma 4.20, since the quantity in the right-hand side does not depend on m, we have

$$\int_{S^{n-1}} \xi_1^2\, d\mu(\xi) = 0,$$

so the measure μ is supported in the intersection of the hyperplane $\xi_1 = 0$ with the sphere S^{n-1}. It follows now from (4.28) and condition (i) that $(\|x\|^{-p})''_{x_1^2}(x) = 0$ for every $x \in \mathbb{R}^n$ with $(x_2, ..., x_n) \neq 0$. Therefore, $\|x\|^{-p} = g_0(x_2, ..., x_n) + x_1 g_1(x_2, ..., x_n)$ for some functions g_0 and g_1 on $\mathbb{R}^{n-1} \setminus \{0\}$. This relation finally leads to a contradiction, because $\|x\|^{-p}$ tends to zero, as $x_1 \to \infty$. We have proved Theorem 4.19. □

Let us show several applications. For normed spaces X and Y and $q \in \mathbb{R}$, $q \ge 1$, the q-sum $(X \oplus Y)_q$ of X and Y is defined as the space of pairs $\{(x, y) : x \in X, y \in Y\}$ with the norm

$$\|(x, y)\| = (\|x\|_X^q + \|y\|_Y^q)^{1/q}.$$

The following result generalizes the fact that the unit balls of the spaces ℓ_q^n, $q > 2$, are not k-intersection bodies for $k < n-3$; see Theorem 4.13.

THEOREM 4.21. *Let $k \in \mathbb{N} \cup \{0\}$, $q > 2$, and let Y be a finite dimensional normed spaces with dim$(Y) \ge k+3$. Then the unit ball of the q-sum of $\mathbb{R}$ and Y is not a k-intersection body.*

PROOF. We need to prove that the q-sum of $\mathbb{R}$ and Y satisfies the conditions of Theorem 4.19. This follows from a straightforward calculation of the derivatives by x of the function $(|x|^q + \|y\|^q)^{1/q}$. □

Another example of a space satisfying the conditions of Theorem 4.19 is a finite dimensional Orlicz space ℓ_M^n, whose Orlicz function $M \in C^2([0,\infty))$ and $M'(0) = M''(0) = 0$. Recall that an *Orlicz function* M is a non-decreasing convex function on $[0,\infty)$ such that $M(0) = 0$ and $M(t) > 0$ for every $t > 0$.

The norm $\|\cdot\|_M$ of the n-dimensional Orlicz space ℓ_M^n is defined implicitly by the equality $\sum_{k=1}^n M(|x_k|/\|x\|_M) = 1,\ x \in \mathbb{R}^n \setminus \{0\}$.

THEOREM 4.22. *Suppose that M is an Orlicz function such that $M \in C^2([0,\infty))$ and $M'(0) = M''(0) = 0$. Then, for every $k < n-3$ the unit ball of the n-dimensional Orlicz space ℓ_M^n is not a k-intersection body.*

PROOF. We are going to show that the norm of the space ℓ_M^n satisfies the conditions of Theorem 4.19. Since the Orlicz norm is an even function with respect to each variable, it suffices to consider the points $x = (x_1,\dots,x_n)$ with non-negative coordinates. We denote by $e_1,\dots,e_n$ the standard normalized basis in ℓ_M^n. We can assume, without loss of generality, that $M(1) = 1$ and, then, $e_1,\dots,e_n$ is the standard basis in $\mathbb{R}^n$. We simply write $\|x\|$ for $\|x\|_M$.

The function M' is non-decreasing, continuous on $[0,\infty)$ and also $M'(0) = 0$. Since $M(0) = 0$ and $M(t) > 0$ for every $t > 0$, the function M' cannot be equal to zero on an interval, so $M'(t) > 0$ for every $t > 0$.

Let $x = (x_1,\dots,x_n)$ with $(x_2,\dots,x_n) \neq 0$. Then $x_m M'(x_m/\|x\|)$ is a positive number for some $2 \le m \le n$. By implicit differentiation,

$$\|x\|'_{x_1} = \frac{\|x\| M'(\frac{x_1}{\|x\|})}{\sum_{m=1}^n x_m M'(\frac{x_m}{\|x\|})}. \tag{4.29}$$

Also,

$$\|x\|''_{x_1^2} = \frac{(\|x\| - x_1\|x\|'_{x_1})^2 M''(\frac{x_1}{\|x\|}) + \sum_{m=2}^n x_m^2 (\|x\|'_{x_1})^2 M''(\frac{x_m}{\|x\|})}{\|x\|^2 \sum_{m=1}^n x_m M'(\frac{x_m}{\|x\|})}. \tag{4.30}$$

Condition (i) of Theorem 4.19 follows from the fact that $M'(0) = M''(0) = 0$.

Let us show that the norm satisfies condition (ii) of Theorem 4.19. Let

$$c = \min\{\sum_{m=2}^n x_m M'(x_m/2) : \|x_2e_2 + \cdots + x_ne_n\| = 1,\ x_2,...,x_n \ge 0\}.$$

Since M' is a continuous function and $M'(t) > 0$ for $t > 0$, we have $c > 0$. Let $d = \max_{t\in[0,1]} M''(t)$.

Clearly, $x_i \le \|x\|,\ i = 1,...,n$. Therefore, also using remark (ii) and positivity of $\|x\|'_{x_1}$, we have $0 \le \|x\| - x_m\|x\|'_{x_1} \le \|x\|$, for every $1 \le m \le n$, and $(\|x\| - x_1\|x\|'_{x_1})^2/\|x\|^2 \le 1$.

Consider any $x_2,...,x_n \ge 0$ with $\|x_2e_2 + \cdots + x_ne_n\| = 1$. Then $x_2,...,x_n \le 1$. If $x_1 \in [0,1]$, then $1 \le \|x\| \le 2$; hence, $x_i/\|x\| \ge (x_i/2),\ i = 1,...,n$. We get from (4.30) that $\|x\|''_{x_1^2} \le nd/c$.

If $x_1 > 1$, then $x_1/\|x\| > 1/2$, and equality (4.30) implies

$$\|x\|''_{x_1^2} \le nd/M'(1/2),$$

because M' is an increasing function.

Since in both cases $x_1 \in [0,1]$ and $x_1 > 1$, we estimated the second derivative by constants which do not depend on the choice of $x_2, ..., x_n$ with $\|x_2e_2+...+x_ne_n\| = 1$, we get condition (ii) of Theorem 4.19.

Finally, let us show condition (iii) of Theorem 4.19. Let c be as defined above. Since M' is a continuous increasing function whose value at zero is zero, for every $\epsilon > 0$ there exists $\delta > 0$ so that $M'(t) < c\epsilon/2$ if $t < \delta$. Then, by (4.29), for every $x_2, ..., x_n$ with $\|x_2e_2 + ... + x_ne_n\| = 1$, we have $\|x\|'_{x_1}(x_1, ..., x_n) \leq \epsilon$ if $x_1 < \min(1, \delta)$, which proves that the first derivative converges to zero uniformly. Similarly, we use (4.30), the uniform convergence of the first derivative and the fact that M'' is continuous and $M''(0) = 0$ to prove that the second derivative of the norm also converges to zero uniformly. □

4.5. Remarks and further results

The concept of an intersection body of a star body was introduced by Lutwak [Lu1] in 1988, as part of his dual Brunn-Minkowski theory. The more general concept of an intersection body first appeared in [GLW]. However, several results related to intersection bodies were proved earlier (without naming this special class of bodies and only considering intersection bodies of convex bodies). For example, Busemann's theorem, Theorem 2.5, shows that if L is an origin-symmetric convex body, then the intersection body of L is also origin-symmetric convex. Another proof and a generalization of this result were found by Milman and Pajor [MiP].

Lutwak [Lu1] established a connection between intersection bodies and the Busemann-Petty problem that reduced the problem to the study of intersection bodies; see Section 5.1. The first example of a non-intersection body was given by Gardner [Ga1] in 1994, as a rounded cylinder in dimension 5. More examples were given by Zhang [Zh2] also in 1994. At the same time, Gardner [Ga2] proved that every symmetric convex body in $\mathbb{R}^3$ is an intersection body using geometric methods.

The Fourier analytic characterization of intersection bodies, Theorem 4.1, was found in [K10]. Several applications of this result appeared immediately after that. It was proved in [K8] that, contrary to the claim in [Zh2], the unit cube in $\mathbb{R}^4$ is an intersection body, which reopened the four-dimensional case of the Busemann-Petty problem; see Section 5.5. Another result of [K10] showed that the unit balls of finite dimensional subspaces of L_p, $0 < p \leq 2$, are intersection bodies; see Theorem 4.11. In particular, every polar projection body (unit ball of a subspace of L_1) is an intersection body which proves a conjecture of M. Meyer [Me]. Using the Fourier criterion, it was proved in [K8] that the unit balls of the spaces ℓ_q^n, $q > 2$, $n \geq 5$, are not intersection bodies, and more examples of non-intersection bodies were given in [K9]; see the second derivative test from Theorem 4.19. After these developments, Zhang [Zh4] proved that every origin-symmetric convex body in $\mathbb{R}^4$ is an intersection body, using methods similar to those in [Ga2], and the same result was proved in [GKS] using Fourier analytic methods. Our proof of Corollary 4.2 is the one from [GKS]. All these results appeared as preprints in May-July 1997 and were published within the next two years.

The concept of a k-intersection body was introduced in [K13] and [K14]. These bodies are related to a generalization of the Busemann-Petty problem from Section 5.2 in the same way that intersection bodies are related to the original Busemann-Petty problem. The Fourier analytic characterization of these bodies, Lemma 4.5,

Theorems 4.6 and 4.8, was proved in [K13] and [K14]. Corollary 4.9 was proved in [K11]. The results of Theorems 4.11, 4.15 and 4.13 were proved in [K10] and [K8] and are formulated in Section 4.3 in the language of k-intersection bodies. The second derivative test, Theorem 4.19, and its applications in Section 4.4 are generalizations of a result from [K9].

Zhang [Zh3] introduced another generalization of intersection bodies. This generalization is related to the lower dimensional version of the Busemann-Petty problem; see a proof of one direction and more details in Section 5.5. For $1 \le k \le n-1$, let $G(n, n-k)$ be the Grassmanian of $(n-k)$-dimensional subspaces of $\mathbb{R}^n$. Recall that the $(n-k)$-dimensional spherical Radon transform is an operator $R_{n-k} : C(S^{n-1}) \mapsto C(G(n, n-k))$ defined by

$$R_{n-k}f(H) = \int_{S^{n-1}\cap H} f(x)\, dx, \quad H \in G(n, n-k).$$

Denote the image of the operator R_{n-k} by X:

$$R_{n-k}(C(S^{n-1})) = X \subset C(G(n, n-k)).$$

Let $M^+(X)$ be the space of linear positive functionals on X, i.e. for every $\mu \in M^+(X)$ and non-negative function $f \in X$ we have $\mu(f) \ge 0$.

An origin-symmetric star body D in $\mathbb{R}^n$ is called a *generalized k-intersection body* if there exists a functional $\mu \in M^+(X)$ so that for every $f \in C(S^{n-1})$,

$$\int_{S^{n-1}} \|x\|_D^{-k} f(x)\, dx = \mu(R_{n-k}f). \tag{4.31}$$

In Section 5.5 we show how the following fact, proved in [K14], implies a negative solution to the lower dimensional Busemann-Petty problem for dimensions of sections greater than 3.

THEOREM 4.23. *Every generalized k-intersection body in $\mathbb{R}^n$ is a k-intersection body.*

PROOF. For any non-negative test function ϕ,

$$\langle (\|\cdot\|_D^{-k})^\wedge, \phi\rangle = \int_{\mathbb{R}^n} \|x\|_D^{-k} \hat{\phi}(x)\, dx$$

$$= \int_{S^{n-1}} \|\theta\|_D^{-k} \left(\int_0^\infty r^{n-k-1}\hat{\phi}(r\theta)\, dr \right) d\theta. \tag{4.32}$$

By (4.31) applied to the function

$$f(\theta) = \int_0^\infty r^{n-k-1}\hat{\phi}(r\theta)\, dr,$$

the expression in (4.32) is equal to

$$\mu\left(R_{n-k}\left(\int_0^\infty r^{n-k-1}\hat{\phi}(r\theta)\, dr \right)\right)$$

$$= \mu\left(\int_{S^{n-1}\cap H} \left(\int_0^\infty r^{n-k-1}\hat{\phi}(r\theta)\, dr \right) d\theta \right)$$

$$= \mu\left(\int_H \hat{\phi} \right) = \mu\left((2\pi)^{n-k} \int_{H^\perp} \phi \right) \ge 0,$$

where we use the equality $\int_H \hat{\phi} = (2\pi)^{n-k} \int_{H^\perp} \phi$ from Lemma 3.24. So $\|\cdot\|_D^{-k}$ is a positive definite distribution, and the result follows from Theorem 4.8. □

It is not known whether the inverse statement is true. If it is true, it would imply an affirmative answer to the lower dimensional Busemann-Petty problem with two- and three-dimensional sections; see Section 5.5.

Intersection bodies are related to L_p-spaces. We discuss this connection in detail in Section 6.3. In Section 6.4 one can find several remarks on isomorphic classification of intersection bodies.

There are many other interesting results related to intersection bodies. We refer the reader to the book [Ga3] for some of them. Let us just mention that Zhang [Zh5] has proved that no convex polytope can be the intersection body of a star body. More results characterizing intersection bodies of star bodies can be found in the papers by Campi [Ca1], [Ca2].

Goodey and Weil [GW2] have proved that all intersection bodies can be obtained as the closure in the radial metric of radial sums of ellipsoids. This implies, in particular, that central sections of intersection bodies of any dimension are themselves intersection bodies in the space of corresponding dimension. The latter was also proved in [FGW]. E. Milman [Mi] used the Fourier characterization of k-intersection bodies to prove the more general fact that a central section of a k-intersection body is also a k-intersection body provided that k is smaller than the dimension of the section.

The paper [GrZ] contains many interesting results on generalized k-intersection bodies. In particular the result of Goodey and Weil can be extended to generalized k-intersection bodies: an origin-symmetric body K in $\mathbb{R}^n$ is a generalized k-intersection body if and only if it is the limit in the radial metric of $(-k)$-sums of ellipsoids (see the definition in Section 2.1). A different proof of this fact was given by E. Milman [Mi], who also noted that this immediately implies the result of Theorem 4.23. The question of whether every k-intersection body is the limit in the radial metric of $(-k)$-sums of ellipsoids is open. An affirmative answer to this question with $k = n-2$ and $k = n-3$ would solve in the affirmative the lower dimensional Busemann-Petty problem for two- and three-dimensional sections; see Section 5.5.

CHAPTER 5

The Busemann-Petty Problem

In this section we consider the following problem posed by Busemann and Petty in 1956: suppose that K and L are origin-symmetric convex bodies in $\mathbb{R}^n$ so that

$$\text{Vol}_{n-1}(K \cap \xi^\perp) \le \text{Vol}_{n-1}(L \cap \xi^\perp), \qquad \forall \xi \in S^{n-1}. \tag{5.1}$$

Does it follow that

$$\text{Vol}_n(K) \le \text{Vol}_n(L)?$$

By association with the Cavalieri principle one can think that the solution is affirmative. And, clearly, it is affirmative in dimension 2, because condition (5.1) means in this case that the radius of K in every direction is smaller than the radius of L, so $K \subset L$. In higher dimensions, however, the situation may be different. In the Cavalieri principle one is allowed to consider all hyperplane sections perpendicular to a given direction, and then the volume of a body can be expressed as the integral of hyperplane volumes of all these sections. In the Busemann-Petty problem only central hyperplane sections are considered, and it makes a big difference. We show in this section that the solution to the Busemann-Petty problem is affirmative if $n \le 4$ and it is negative if $n \ge 5$. The history of the solution is presented in the Introduction, and related open problems are discussed in Section 5.5.

5.1. A Fourier analytic solution

The Fourier analytic solution to the Busemann-Petty problem is based on three main ingredients. The first is a connection between the Busemann-Petty problem and intersection bodies found by Lutwak [Lu1], the second is the connection between intersection bodies and the Fourier transform, Theorem 4.1, and the third is the formula relating the derivatives of the parallel section function to the Fourier transform of powers of the norm, Theorem 3.18. We have the last two tools ready, so we need to prove Lutwak's connection:

THEOREM 5.1. *The solution to the Busemann-Petty problem in $\mathbb{R}^n$ is affirmative if and only if every origin-symmetric convex body in $\mathbb{R}^n$ is an intersection body.*

We present a geometric proof that consist of two parts.

THEOREM 5.2. *If K is an origin-symmetric star body in $\mathbb{R}^n$ that is an intersection body, then for any origin-symmetric star body L satisfying (5.1), we have $\text{Vol}_n(K) \le \text{Vol}_n(L)$, i.e. the solution to the Busemann-Petty problem for K and L is affirmative.*

PROOF. Using the expression for the volume of sections in terms of the spherical Radon transform (2.17), we see that condition (5.1) is equivalent to

$$R(\|\cdot\|_K^{-n+1})(\xi) \le R(\|\cdot\|_L^{-n+1})(\xi), \quad \forall \xi \in S^{n-1}. \tag{5.2}$$

Since K is an intersection body, there exists a finite measure μ on S^{n-1} so that $\|\cdot\|_K^{-1}$ and $R\mu$ are equal functionals on $C(S^{n-1})$. Applying both functionals to a continuous function $\|\cdot\|_L^{-n+1}$, we get

$$\int_{S^{n-1}} \|x\|_K^{-1}\|x\|_L^{-n+1}dx = (\|\cdot\|_K^{-1}, \|\cdot\|_L^{-n+1})$$

$$= (R\mu, \|\cdot\|_L^{-n+1}) = (\mu, R(\|\cdot\|_L^{-n+1})) = \int_{S^{n-1}} R(\|\cdot\|_L^{-n+1})(\xi)d\mu(\xi).$$

Using (5.2), positivity of the measure and repeating the above calculation in the opposite order, we get that the latter quantity is greater than or equal to

$$\int_{S^{n-1}} R(\|\cdot\|_K^{-n+1})(\xi)d\mu(\xi) = \int_{S^{n-1}} \|x\|_K^{-1}\|x\|_K^{-n+1}dx = n\mathrm{Vol}_n(K).$$

The last equality here is the polar formula for volume, formula (2.3). Hence, by Hölder's inequality,

$$n\mathrm{Vol}_n(K) \le \int_{S^{n-1}} \|x\|_K^{-1}\|x\|_L^{-n+1}dx$$

$$\le \left(\int_{S^{n-1}} \|x\|_K^{-n}dx\right)^{1/n} \left(\int_{S^{n-1}} \|x\|_L^{-n}dx\right)^{(n-1)/n}$$

$$= n\left(\mathrm{Vol}_n(K)\right)^{1/n}\left(\mathrm{Vol}_n(L)\right)^{(n-1)/n},$$

which implies $\mathrm{Vol}_n(K) \le \mathrm{Vol}_n(L)$. □

We are going to use a standard perturbation procedure for convex bodies. Namely, if K is a 2-smooth convex body with strictly positive curvature, $p \in \mathbb{R}$ is a non-zero number, and f is any C^2 function on the sphere, then for small enough ϵ the equality

$$\|x\|_L^p = \|x\|_K^p + \epsilon f(x/|x|_2)|x|_2^p$$

defines a convex body L. This essentially follows from a simple two-dimensional argument: if h is a strictly concave function on a bounded interval so that $h'' < \delta < 0$, and g is a twice differentiable function on this interval, then for small ϵ the second derivative of the function $h + \epsilon g$ is still strictly negative on the whole interval, so $f + \epsilon g$ is also concave. A more general fact can be found in Lemma 5.16.

THEOREM 5.3. *If there exists an origin-symmetric convex body in $\mathbb{R}^n$ that is not an intersection body, then one can construct two origin-symmetric convex bodies K and L in $\mathbb{R}^n$ providing a counterexample to the Busemann-Petty problem, i.e. for every $\xi \in S^{n-1}$,*

$$\mathrm{Vol}_{n-1}(K \cap \xi^\perp) \le \mathrm{Vol}_{n-1}(L \cap \xi^\perp),$$

but

$$\mathrm{Vol}_n(K) > \mathrm{Vol}_n(L).$$

PROOF. First, by the approximation argument of Lemma 4.10, there exists an infinitely smooth, origin-symmetric convex body L in $\mathbb{R}^n$ with strictly positive curvature that is not an intersection body. The function $(1/(2^n\pi^{n-1}))\|\cdot\|_L^{-1}$ satisfies the conditions of the second statement of Lemma 3.16, so its Fourier transform has the form $g_0(\theta)r^{-n+1}$, where $g_0 \in C^\infty(S^{n-1})$. Since $g_0(\theta)r^{-n+1}$ is a homogeneous function of degree $-n+1$, by Lemma 3.7, the spherical Radon transform of $g = (1/(2\pi)^n)g_0$ is equal to $\|\cdot\|_L^{-1}$, i.e. $Rg = \|\cdot\|_L^{-1}$ on the sphere.

The function g is sign-changing, because otherwise L would be an intersection body. Let $\Omega \subset S^{n-1}$ be an open set where $g < 0$. Consider a function $h \in C^\infty(S^{n-1})$ so that $h \geq 0$ everywhere on S^{n-1} and $h > 0$ only on a non-empty subset of Ω. Extend h to a homogeneous function of degree -1 on $\mathbb{R}^n$. By the last statement of Lemma 3.16, the Fourier transform of $h(\theta)r^{-1}$ is a function $(1/(2\pi)^n)u(\theta)r^{-n+1}$, where $u \in C^\infty(S^{n-1})$. By Lemma 3.7, $h = Ru$.

Now we are ready to construct the body K. Define K by

$$\|\theta\|_K^{-n+1} = \|\theta\|_L^{-n+1} - \epsilon u, \quad \theta \in S^{n-1}.$$

Since L has strictly positive curvature, by the argument preceding this theorem, one can choose a small enough ϵ so that the body K is convex.

Next, for every $\xi \in S^{n-1}$, since $Ru = h \geq 0$,

$$R(\|\cdot\|_K^{-n+1})(\xi) = R(\|\cdot\|_L^{-n+1})(\xi) - \epsilon h(\xi) \leq R(\|\cdot\|_L^{-n+1})(\xi),$$

which means that K and L satisfy condition (5.2), so

$$\mathrm{Vol}_{n-1}(K \cap \xi^\perp) \leq \mathrm{Vol}_{n-1}(L \cap \xi^\perp), \qquad \forall \xi \in S^{n-1}.$$

On the other hand, by self-duality of the spherical Radon transform, Lemma 2.13, and the fact that the product hg is non-positive everywhere, we have (recall that $Rg = \|\cdot\|_L^{-1}$)

$$\int_{S^{n-1}} \|\theta\|_K^{-n+1}\|\theta\|_L^{-1}d\theta = \int_{S^{n-1}} \|\theta\|_K^{-n+1}Rg(\theta)d\theta$$

$$= \int_{S^{n-1}} R(\|\cdot\|_K^{-n+1})(\xi)g(\xi)d\xi$$

$$= \int_{S^{n-1}} R(\|\cdot\|_L^{-n+1})(\xi)g(\xi)d\xi - \epsilon\int_{S^{n-1}} h(\xi)g(\xi)d\xi$$

$$> \int_{S^{n-1}} R(\|\cdot\|_L^{-n+1})(\xi)g(\xi)d\xi = \int_{S^{n-1}} \|\theta\|_L^{-n+1}Rg(\theta)d\theta$$

$$= \int_{S^{n-1}} \|\theta\|_L^{-n+1}\|\theta\|_L^{-1}d\theta = n\mathrm{Vol}_n(L).$$

Applying Hölder's inequality in the same way as in Theorem 5.2, we get $\mathrm{Vol}_n(K) > \mathrm{Vol}_n(L)$. □

We are now ready to present the solution of the Busemann-Petty problem. In the proof, besides Lutwak's connection, we use two results about intersection bodies – Corollary 4.9 and Theorem 4.13. Recall that both these results follow from the formula of Theorem 3.18 and the Fourier characterization of intersection bodies in Theorem 4.1. Instead of the argument with ℓ_q^n in the proof below, we could use Corollary 4.4.

THEOREM 5.4. *The solution to the Busemann-Petty problem is affirmative if $n \leq 4$ and it is negative if $n \geq 5$.*

PROOF. By Corollary 4.9, every origin-symmetric convex body in $\mathbb{R}^n$, $n = 2, 3, 4$, is a 1-intersection body (or simply an intersection body). The affirmative part follows now by Theorem 5.2.

Let $n \geq 5$. By Theorem 4.13 (with $k = 1$), for any $q > 2$ the unit ball of the space ℓ_q^n is not an intersection body. The negative part follows from Theorem 5.3. □

A geometric explanation of the transition between the dimensions 4 and 5 is given before Corollary 4.4 and is based on the fact that convexity controls differential properties of the body only up to the second derivatives.

5.2. How can one make the answer affirmative?

In this section we generalize the solution to the Busemann-Petty problem. Since the answer to the original problem is negative in most dimensions, it is natural to ask what condition on the parallel section functions at zero allows us to compare the volumes. We present one possible result of this type. Another feature of this result is that it generalizes the solution of the original problem, but the proof is completely Fourier analytic. The main tool is Parseval's formula on the sphere from Section 3.4.

The next two theorems generalize Lutwak's connections, Theorems 5.2 and 5.3.

THEOREM 5.5. *Let $1 \leq k < n$, and let K and L be origin-symmetric $(k-1)$-smooth star bodies in $\mathbb{R}^n$ if k is odd and k-smooth if k is even. Suppose that the functions*

$$\|\cdot\|_K^{-k} \quad \text{and} \quad \|\cdot\|_L^{-n+k} - \|\cdot\|_K^{-n+k} \tag{5.3}$$

represent positive definite distributions in $\mathbb{R}^n$. Then

$$\mathrm{Vol}_n(K) \leq \mathrm{Vol}_n(L).$$

PROOF. The result of Corollary 3.17 implies that the distributions $(\|\cdot\|_K^{-n+k})^\wedge$ and $(\|\cdot\|_K^{-n+k})^\wedge$ are continuous functions outside of the origin, so positive definiteness of the second function in (5.3) can be written in the form

$$(\|\cdot\|_L^{-n+k})^\wedge(\theta) \geq (\|\cdot\|_K^{-n+k})^\wedge(\theta), \qquad \forall \theta \in S^{n-1}. \tag{5.4}$$

Let μ_0 be the measure on S^{n-1} that corresponds to a positive definite distribution $\|\cdot\|_K^{-k}$ by Corollary 2.26. By the Parseval formula on the sphere, Corollary 3.23,

$$\int_{S^{n-1}} (\|\cdot\|_L^{-n+k})^\wedge(\theta) d\mu_0(\theta) = \int_{S^{n-1}} \|\theta\|_L^{-n+k} \|\theta\|_K^{-k} d\theta.$$

At the same time,

$$\int_{S^{n-1}} (\|\cdot\|_K^{-n+k})^\wedge(\theta) d\mu_0(\theta) = \int_{S^{n-1}} \|\theta\|_K^{-n+k} \|\theta\|_K^{-k} d\theta.$$

Using (5.4) and positivity of μ_0, we see that the left-hand side of the first equation is greater than the left-hand side of the second equation. Therefore, the right-hand sides also compare as

$$n\mathrm{Vol}_n(K) = \int_{S^{n-1}} \|\theta\|_K^{-n+k} \|\theta\|_K^{-k} d\theta \leq \int_{S^{n-1}} \|\theta\|_L^{-n+k} \|\theta\|_K^{-k} d\theta.$$

By Hölder's inequality, we estimate the right-hand side by

$$\leq \left(\int_{S^{n-1}} \|\theta\|_K^{-n} d\theta \right)^{k/n} \left(\int_{S^{n-1}} \|\theta\|_L^{-n} d\theta \right)^{(n-k)/n}$$
$$= n \left(\mathrm{Vol}_n(K)\right)^{k/n} \left(\mathrm{Vol}_n(L)\right)^{(n-k)/n}.$$

We get that

$$\mathrm{Vol}_n(K) \le (\mathrm{Vol}_n(K))^{k/n} (\mathrm{Vol}_n(L))^{(n-k)/n},$$

which implies $\mathrm{Vol}_n(K) \le \mathrm{Vol}_n(L)$. □

THEOREM 5.6. *Let $1 \le k < n$, and suppose there exists an origin-symmetric convex body D in $\mathbb{R}^n$ for which $\|\cdot\|_D^{-k}$ is not a positive definite distribution. Then there exist origin-symmetric infinitely smooth convex bodies K and L so that*

$$\|\cdot\|_L^{-n+k} - \|\cdot\|_K^{-n+k}$$

is a positive definite distribution, but

$$\mathrm{Vol}_n(K) > \mathrm{Vol}_n(L).$$

PROOF. By Theorem 4.8, the body D is not a k-intersection body. By Lemma 4.10, there exists an origin-symmetric infinitely smooth convex body L with strictly positive curvature, which also is not a k-intersection body. By Theorem 4.8, the latter means that the continuous (by Corollary 3.17) function $(\|\cdot\|_L^{-k})^\wedge$ is negative on some open set $\Omega \subset S^{n-1}$.

Let $f \in C^\infty(S^{n-1})$ be a non-negative even function supported in Ω. Extend f to a homogeneous function $f(\theta)r^{-k}$ of degree $-k$ on $\mathbb{R}^n$. By Lemma 3.16, the Fourier transform of $f(\theta)r^{-k}$ is a homogeneous function of degree $-n+k$ on $\mathbb{R}^n$, whose restriction to the sphere is infinitely smooth: $(f(\theta)r^{-k})^\wedge = g(\theta)r^{-n+k}$, where $g \in C^\infty(S^{n-1})$. Define a body K by

$$\|x\|_K^{-n+k} = \|x\|_L^{-n+k} - \epsilon g(\theta)r^{-n+k}.$$

Since L is infinitely smooth, convex and has positive curvature, by the argument preceding Theorem 5.3, one can choose ϵ small enough so that the body K is convex. Also K is infinitely smooth. By the definition of g and since f is non-negative,

$$(\|\cdot\|_K^{-n+k})^\wedge = (\|\cdot\|_L^{-n+k})^\wedge - (2\pi)^n \epsilon f(\theta)r^{-k} \le (\|\cdot\|_L^{-n+k})^\wedge,$$

so the distribution $\|x\|_L^{-n+k} - \|x\|_K^{-n+k}$ is positive definite. On the other hand, the function f is positive only where $(\|\cdot\|_L^{-k})^\wedge$ is negative, so

$$\int_{S^{n-1}} (\|\cdot\|_K^{-n+k})^\wedge(\xi)(\|\cdot\|_L^{-k})^\wedge(\xi)d\xi$$

$$= \int_{S^{n-1}} (\|\cdot\|_L^{-n+k})^\wedge(\xi)(\|\cdot\|_L^{-k})^\wedge(\xi)d\xi - \epsilon \int_{S^{n-1}} (\|\cdot\|_L^{-k})^\wedge(\xi)f(\xi)d\xi$$

$$> \int_{S^{n-1}} (\|\cdot\|_L^{-n+k})^\wedge(\xi)(\|\cdot\|_L^{-k})^\wedge(\xi)d\xi.$$

Now apply Parseval's formula in the function form, Lemma 3.22:

$$\int_{S^{n-1}} \|x\|_L^{-n+k}\|x\|_L^{-k}dx < \int_{S^{n-1}} \|x\|_K^{-n+k}\|x\|_L^{-k}dx.$$

Using Hölder's inequality in the same way, as in Theorem 5.5, we get $\mathrm{Vol}_n(K) > \mathrm{Vol}_n(L)$. □

The next result shows that, in order to compare volumes, one needs information about the $(n-4)$-th derivatives of parallel section functions at zero. It is not enough to have inequalities for the $(n-5)$-th derivatives.

THEOREM 5.7. *Let K and L be origin-symmetric $(k-1)$-smooth convex bodies in $\mathbb{R}^n$, where k is an odd integer, $0 < k < n$. Suppose that for every $\xi \in S^{n-1}$,*

$$(-1)^{(k-1)/2} A_{K,\xi}^{(k-1)}(0) \le (-1)^{(k-1)/2} A_{L,\xi}^{(k-1)}(0). \tag{5.5}$$

Then the following hold.
(i) If $k \ge n-3$, we have $\mathrm{Vol}_n(K) \le \mathrm{Vol}_n(L)$.
(ii) If $k < n-3$, it is still possible that $\mathrm{Vol}_n(K) > \mathrm{Vol}_n(L)$.

PROOF. Note that, by Theorem 3.18, condition (5.5) is equivalent to the positive definiteness of $\|\cdot\|_L^{-n+k} - \|\cdot\|_K^{-n+k}$.

If $k \ge n-3$, then, by Corollary 4.9, the function $\|\cdot\|_K^{-k}$ represents a positive definite distribution, and the first statement of the theorem follows from Theorem 5.5.

If $k < n-3$, then, by Theorem 4.13, the unit balls of the spaces ℓ_q^n, $q > 2$, have the property that the distribution $\|\cdot\|_q^{-k}$ is not positive definite, and we can use any of these balls as the body D in Theorem 5.6. The result follows. □

Theorem 5.7 holds in the case where k is even, but condition (5.5) must be replaced by an inequality for the expressions appearing in the second formula of Theorem 3.18.

The statements of Theorems 5.5 and 5.6 can be translated into the language of geometry, where they directly generalize Lutwak's connections between intersection bodies and the Busemann-Petty problem. Just note that the condition that $\|\cdot\|_K^{-k}$ is positive definite means that K is a k-intersection body, and the condition that $\|\cdot\|_L^{-n+k} - \|\cdot\|_K^{-n+k}$ is a positive definite distribution is equivalent to (5.5).

5.3. The affirmative part via spherical harmonics

Historically, spherical harmonics are the most popular Fourier analytic tool in convex geometry. The book by Groemer [Gr] contains many interesting geometric applications of spherical harmonics, historical remarks and references. In this section we present a proof of the affirmative part of the Busemann-Petty problem based on the Funk-Hecke formula for spherical harmonics.

Let us start with a few facts about spherical harmonics. We denote by P_m the space of spherical harmonics of degree m on the unit sphere S^{n-1} in $\mathbb{R}^n$. Recall that spherical harmonics of degree m are restrictions to the sphere of harmonic homogeneous polynomials of degree m. We consider P_m as a subspace of $L_2(S^{n-1})$. Any two spherical harmonics of different degrees are orthogonal in $L_2(S^{n-1})$ [Mu, p. 2]

The dimension $N(n,m)$ of the space P_m can easily be calculated [Mu, p. 4]:

$$N(n,m) = \frac{(2m+n-2)\Gamma(n+m-2)}{\Gamma(m+1)\Gamma(n-1)}. \tag{5.6}$$

Let $\{Y_{mj} : j = 1, ..., N(n,m)\}$ be an orthonormal basis of the space P_m. The functions Y_{mj}, $m \in \mathbb{N} \cup \{0\}$, $j = 1, ..., N(n,m)$, form an orthonormal basis of the space $L_2(S^{n-1})$, [Gr, p. 71]. Therefore, if $F, G \in L_2(S^{n-1})$, then

$$(F,G) = \sum_{m=0}^{\infty} \sum_{j=1}^{N(n,m)} (F, Y_{mj})(G, Y_{mj}),$$

where (F,G) stands for the scalar product in $L_2(S^{n-1})$. Also,

$$\|F\|^2_{L_2(S^{n-1})} = \sum_{m=0}^{\infty} \sum_{j=1}^{N(n,m)} (F, Y_{mj})^2.$$

An important role in this section belongs to the Funk-Hecke formula [Mu, p. 20] or [Gr, p. 98]: for every $Y_m \in P_m$, every continuous function f on $[-1,1]$ and $x \in S^{n-1}$,

$$\int_{S^{n-1}} f((x,\xi))Y_m(\xi)d\xi = \lambda_m Y_m(x), \tag{5.7}$$

where

$$\lambda_m = \frac{(-1)^m \pi^{(n-1)/2}}{2^{m-1}\Gamma(m+\frac{n-1}{2})} \int_{-1}^{1} f(t) \frac{d^m}{dt^m}(1-t^2)^{m+\frac{n-3}{2}}\, dt. \tag{5.8}$$

Let us calculate the coefficients λ_m in the case where $f(t) = |t|^q, q > -1$. We denote these coefficients by $\lambda_m(q)$. Clearly, $\lambda_m(q) = 0$ if m is an odd integer.

LEMMA 5.8. *Let $q > -1,\ q \neq 2k,\ k \in \mathbb{N} \cup \{0\}$. Then for every even integer $m \geq 0$,*

$$\lambda_m(q) = \frac{\pi^{n/2-1}\Gamma(q+1)\sin(\pi(m-q)/2)\Gamma((m-q)/2)}{2^{q-1}\Gamma((m+n+q)/2)}. \tag{5.9}$$

PROOF. First, let $q > m$ and calculate the integral from (5.8) by parts m times:

$$\lambda_m(q) = \frac{(-1)^m \pi^{(n-1)/2} q(q-1)...(q-m+1)}{2^{m-1}\Gamma(m+(n-1)/2)}$$
$$\times \int_{-1}^{1} |t|^{q-m}(1-t^2)^{m+\frac{n-3}{2}}\, dt.$$

Now use formulas (2.31), (2.29) and (2.32) to get (5.9) with $q > m$.

To prove (5.9) for other values of q, we use analytic continuation. First, put $f(t) = |t|^q$ in the left-hand side of (5.8) and use the same argument as in Lemma 2.22 (direct differentiation by q). We get that $\lambda_m(q)$ is an analytic function of q in the domain $\Re q > -1$. The expression in the right-hand side of (5.9) is an analytic function of q in the same domain, except for even integers, at which $\Gamma((m-q)/2)$ may have a pole. By the uniqueness of analytic continuation, (5.9) holds for every q from the required domain. □

By Theorem 2.10, to prove the affirmative part of the Busemann-Petty problem, it is enough to consider the case where K and L are infinitely smooth bodies. In fact, we can approximate K in the radial metric by infinitely smooth convex bodies contained in K and then do the same for L with bodies containing L.

As before, $A^{(q)}_{K,\xi}(0)$ stands for the fractional derivative of the order q at zero of the parallel section function. By the remark preceding Theorem 3.18, $A^{(q)}_{K,\xi}(0)$ is a continuous function of $\xi \in S^{n-1}$. Let us compute the coefficients of the expansion of this function with respect to the basis of spherical harmonics in $L_2(S^{n-1})$.

LEMMA 5.9. *Let K be an infinitely smooth origin-symmetric convex body in $\mathbb{R}^n$. For every $q \in \mathbb{C}$, $-1 < \Re(q) < n-1$, $m = 0, 2, 4, \ldots$, and any spherical harmonic $Y_m \in P_m$ of degree m,*

$$\int_{S^{n-1}} A_{K,\xi}^{(q)}(0) Y_m(\xi)\, d\xi \tag{5.10}$$

$$= \frac{2^{q+1}\pi^{n/2-1}\sin(\pi(m+q+1)/2)\Gamma((m+q+1)/2)}{(n-q-1)\Gamma((m+n-q-1)/2)}$$

$$\times \int_{S^{n-1}} \|\theta\|_K^{-n+q+1} Y_m(\theta)\, d\theta.$$

PROOF. Assume that $q \in (-1, 0)$. Then, by the Fubini theorem,

$$A_{K,\xi}^{(q)}(0) = \frac{1}{\Gamma(-q)} \int_0^\infty t^{-1-q} A_{K,\xi}(t)\, dt$$

$$= \frac{1}{2\Gamma(-q)} \int_{-\infty}^{\infty} |t|^{-1-q} \left(\int_{(x,\xi)=t} \chi(\|x\|_K)\, dx \right) dt$$

$$= \frac{1}{2\Gamma(-q)} \int_K |(x,\xi)|^{-1-q}\, dx.$$

By the Funk-Hecke formula (see (5.7)) with $f(t) = |t|^{-1-q}$ and $\lambda_m = \lambda_m(-1-q)$,

$$\int_{S^{n-1}} A_{K,\xi}^{(q)}(0) Y_m(\xi)\, d\xi$$

$$= \frac{1}{2\Gamma(-q)} \int_K |x|_2^{-1-q} \left(\int_{S^{n-1}} |(\frac{x}{|x|_2}, \xi)|^{-1-q} Y_m(\xi)\, d\xi \right) dx$$

$$= \frac{1}{2\Gamma(-q)} \lambda_m(-1-q) \int_K |x|_2^{-1-q} Y_m(\frac{x}{|x|_2})\, dx$$

$$= \frac{1}{2\Gamma(-q)} \lambda_m(-1-q) \int_{\mathbb{R}^n} |x|_2^{-1-q} \chi(\|x\|_K) Y_m(\frac{x}{|x|_2})\, dx$$

$$= \frac{1}{2\Gamma(-q)} \lambda_m(-1-q) \int_{S^{n-1}} Y_m(\theta) \left(\int_0^\infty r^{n-q-2} \chi(r\|\theta\|_K)\, dr \right) d\theta$$

$$= \frac{1}{2\Gamma(-q)(n-q-1)} \lambda_m(-1-q) \int_{S^{n-1}} \|\theta\|_K^{-n+q+1} Y_m(\theta)\, d\theta.$$

Using the expression for $\lambda_m(-1-q)$ from Lemma 5.9, we get the result for $-1 < q < 0$. By the argument of Lemma 2.22, both sides of (5.10) are analytic functions of q in the domain $\{q \in \mathbb{C} : -1 < \Re(q) < n-1\}$, so the general statement follows by analytic continuation. □

Note that if we replace q by $n-q-2$ in Lemma 5.9, the coefficient essentially reverses. This is a crucial point of the argument. The next lemma shows how this works. As before, we denote by Y_{mj}, $m \in \mathbb{N} \cup \{0\}$, $j = 1, \ldots, N(n,m)$, the orthonormal basis of spherical harmonics in $L_2(S^{n-1})$.

LEMMA 5.10. *Let K, L be infinitely smooth origin-symmetric convex bodies in $\mathbb{R}^n$. Then, for every $q \in \mathbb{C}$, $-1 < \Re(q) < n-1$, we have*

$$\int_{S^{n-1}} A_{K,\xi}^{(n-q-2)}(0) A_{L,\xi}^{(q)}(0)\, d\xi \tag{5.11}$$

$$= \frac{2^n \pi^{n-2} \cos(\pi q/2)\cos(\pi(n-q-2)/2)}{(q+1)(n-q-1)} \int_{S^{n-1}} \|\theta\|_K^{-1-q} \|\theta\|_L^{-n+q+1}\, d\theta.$$

PROOF. Since K and L are origin-symmetric infinitely smooth convex bodies, by the remark preceding Theorem 3.18, $A_{K,\xi}^{(n-q-2)}(0)$ and $A_{L,\xi}^{(q)}(0)$ are even continuous functions of the variable $\xi \in S^{n-1}$. Because of parity, there are only even numbered terms in their spherical harmonics expansions in $L_2(S^{n-1})$ with respect to the basis Y_{mj}. For every even integer $m \geq 0$, we can apply the formula of Lemma 5.9, first to the body L and the number q and then to the body K and the number $n-q-2$ in place of q : for each $j = 1, ..., N(n,m)$,

$$\int_{S^{n-1}} A_{L,\theta}^{(q)}(0)\, Y_{mj}(\theta)\, d\theta \int_{S^{n-1}} A_{K,\theta}^{(n-q-2)}(0)\, Y_{mj}(\theta)\, d\theta$$

$$= \frac{2^n \pi^{n-2} \cos(\pi q/2)\cos(\pi(n-q-2)/2)}{(q+1)(n-q-1)}$$

$$\times \int_{S^{n-1}} \|\theta\|_L^{-n+q+1} Y_{mj}(\theta)\, d\theta \int_{S^{n-1}} \|\theta\|_K^{-q-1} Y_{mj}(\theta)\, d\theta.$$

Note that the coefficient does not depend on m or j. Since the functions Y_{mj} form an orthonormal basis in $L_2(S^{n-1})$, we have

$$\int_{S^{n-1}} A_{K,\xi}^{(n-q-2)}(0) A_{L,\xi}^{(q)}(0)\, d\xi$$

$$= \sum_{m=0}^{\infty} \sum_{j=1}^{N(n,m)} \int_{S^{n-1}} A_{L,\theta}^{(q)}(0)\, Y_{mj}(\theta)\, d\theta \int_{S^{n-1}} A_{K,\theta}^{(n-q-2)}(0)\, Y_{mj}(\theta)\, d\theta$$

$$= \frac{2^n \pi^{n-2} \cos(\pi q/2)\cos(\pi(n-q-2)/2)}{(q+1)(n-q-1)}$$

$$\times \sum_{m=0}^{\infty} \sum_{j=1}^{N(n,m)} \int_{S^{n-1}} \|\theta\|_L^{-n+q+1} Y_{mj}(\theta)\, d\theta \int_{S^{n-1}} \|\theta\|_K^{-q-1} Y_{mj}(\theta)\, d\theta$$

$$= \frac{2^n \pi^{n-2} \cos(\pi q/2)\cos(\pi(n-q-2)/2)}{(q+1)(n-q-1)} \int_{S^{n-1}} \|\theta\|_K^{-1-q} \|\theta\|_L^{-n+q+1}\, d\theta.$$

□

Lemma 5.10 immediately implies the affirmative part of the Busemann-Petty problem.

THEOREM 5.11. *The answer to the Busemann-Petty problem in $\mathbb{R}^4$ is affirmative.*

PROOF. Recall that we can assume without loss of generality that the origin-symmetric convex bodies K and L in the formulation of the Busemann-Petty problem are infinitely smooth. Put $n = 4$ and $q = 0$ in (5.11). We get

$$\int_{S^3} A''_{K,\xi}(0) A_{L,\xi}(0)\, d\xi = -\frac{16\pi^2}{3} \int_{S^3} \|\theta\|_K^{-1} \|\theta\|_L^{-3}\, d\theta. \tag{5.12}$$

Replacing L by K in the latter equality, we get

$$\int_{S^3} A''_{K,\xi}(0)A_{K,\xi}(0)\,d\xi = -\frac{16\pi^2}{3}\int_{S^3}\|\theta\|_K^{-4}\,d\theta. \tag{5.13}$$

By the assumption of the Busemann-Petty problem, for every $\xi \in S^{n-1}$,

$$A_{K,\xi}(0) = \mathrm{Vol}_{n-1}(K\cap\xi^\perp) \le \mathrm{Vol}_{n-1}(L\cap\xi^\perp) = A_{L,\xi}(0).$$

By Brunn's theorem, Theorem 2.3, $A''_{K,\xi}(0) \le 0$ for every ξ. Therefore, the integral in the left-hand side of (5.12) is greater than the integral in the left-hand side of (5.13). Using the polar formula for the volume and Hölder's inequality, we get

$$\begin{aligned} n\mathrm{Vol}_n(K) &= \int_{S^3}\|\theta\|_K^{-4}\,d\theta \le \int_{S^3}\|\theta\|_K^{-1}\|\theta\|_L^{-3}\,d\theta \\ &\le \left(\int_{S^3}\|\theta\|_K^{-4}\,d\theta\right)^{1/4}\left(\int_{S^3}\|\theta\|_L^{-4}\,d\theta\right)^{3/4} \\ &= (n\mathrm{Vol}_n(K))^{1/4}(n\mathrm{Vol}_n(L))^{3/4}. \end{aligned}$$

Clearly, $\mathrm{Vol}_n(K) \le \mathrm{Vol}_n(L)$. □

In the same way, one can prove a more general result that is stronger than the affirmative part of Theorem 5.7.

THEOREM 5.12. *Let K and L be infinitely smooth, origin-symmetric convex bodies in $\mathbb{R}^n$, $n \ge 2$, and let $q \in [n-4, n-1)$, $q > -1$, where q is not an odd integer. Suppose that for every $\xi \in S^{n-1}$,*

$$\frac{A^{(q)}_{K,\xi}(0)}{\cos(\pi q/2)} \le \frac{A^{(q)}_{L,\xi}(0)}{\cos(\pi q/2)}. \tag{5.14}$$

Then $\mathrm{Vol}_n(K) \le \mathrm{Vol}_n(L)$.

PROOF. To prove the result for $q \ne n-3$, it is enough to show that, for every $\xi \in S^{n-1}$,

$$\frac{A^{(n-q-2)}_{K,\xi}(0)}{\cos(\pi(n-q-2)/2)} \ge 0. \tag{5.15}$$

Then our claim follows from Lemma 5.10 in exactly the same way as in Theorem 5.11.

Let $\beta = n-q-2$. Then $\beta \in (-1,2]$, $\beta \ne 1$. The case $\beta = 0$ is trivial, and the case $\beta = 2$ follows from Brunn's theorem, as in the proof of Theorem 5.11.

If $\beta \in (-1,0)$, then

$$A^{(\beta)}_{K,\xi}(0) = \frac{1}{\Gamma(-\beta)}\int_0^\infty t^{-1-\beta}A_{K,\xi}(t)\,dt > 0,$$

and also $\cos(\pi\beta/2) > 0$.

If $\beta \in (0,2)$, $\beta \ne 1$, then it follows from the definition of the fractional derivative (formula (2.42), use the fact that $A_{K,\xi}$ is an even function) that

$$A^{(\beta)}_{K,\xi}(0) = \frac{1}{\Gamma(-\beta)}\int_0^\infty t^{-1-\beta}(A_{K,\xi}(t) - A_{K,\xi}(0))\,dt. \tag{5.16}$$

By Brunn's theorem, for every t, $A_{K,\xi}(t) \le A_{K,\xi}(0)$. On the other hand, the numbers $\cos(\pi\beta/2)$ and $\Gamma(-\beta)$ have opposite signs, which proves our claim.

Now we consider the case $q = n-3$. In this case both sides of formula (5.11) vanish, because K is origin-symmetric and $A_{K,\xi}$ is an even function. Divide both sides of (5.11) by $\cos(\pi(n-q-2)/2)$, and let $q \to n-3$. By (5.16) and Brunn's theorem,

$$\lim_{q\to n-3} \frac{A_{K,\xi}^{(n-q-2)}(0)}{\cos(\pi(n-q-2)/2)} = -\frac{\pi}{2}\int_0^\infty t^{-2}(A_{K,\xi}(t) - A_{K,\xi}(0))\,dt \geq 0.$$

The details of this argument are the same as in the proof of Theorem 3.18.

Now to prove the theorem, divide both sides of (5.11) by both cosines, send q to $n-3$ and argue in the same way as in the proof of Theorem 5.11. □

Remarks. (i) If $q \in [n-4, n-1)$, where $q > -1$ is an odd integer, the result of Theorem 5.12 holds if one replaces condition (5.14) by the condition that, for every $\xi \in S^{n-1}$, the quantity

$$(-1)^{(q+1)/2}\int_0^\infty t^{-q-1}\left(A_{K,\xi}(t) - \sum_{j=0}^{(q-1)/2} \frac{t^{2j}}{(2j)!}A_{K,\xi}^{(2j)}(0)\right)dt$$

is less than or equal to the same expression for the body L. This follows from the same limit argument as in the case $q = n-3$ of Theorem 5.12.

(ii) Formula (5.11) can be written in a different form using formula (3.13) from Theorem 3.18. Then (5.11) becomes Parseval's formula on the sphere (compare it with Lemma 3.22):

$$\int_{S^{n-1}} (\|\cdot\|_K^{-1-q})^\wedge(\xi)(\|\cdot\|_L^{-n+q+1})^\wedge(\xi)\,d\xi$$

$$= (2\pi)^n \int_{S^{n-1}} \|\theta\|_K^{-1-q}\|\theta\|_L^{-n+q+1}\,d\theta.$$

(iii) By Theorems 3.18 and 4.8, the condition

$$\frac{A_{K,\xi}^{(n-q-2)}}{\cos(\pi(n-q-2)/2)} \geq 0, \quad \forall \xi \in S^{n-1},$$

means that K is a fractional analog of a k-intersection body. One can define q-intersection bodies with the non-integer q and generalize Lutwak's connections to this case.

5.4. Zvavitch's generalization to arbitrary measures

Zvavitch [Zv2] found a remarkable generalization of the Busemann-Petty problem, namely, one can replace the volume by essentially any measure on $\mathbb{R}^n$. The answer remains the same, affirmative for $n \leq 4$ and negative for $n \geq 5$.

Let f be an even continuous positive function on $\mathbb{R}^n$, and denote by μ the measure on $\mathbb{R}^n$ with density f.

For every closed bounded set $B \in \xi^\perp$ define

$$\mu(B) = \int_B f(x)dx.$$

We consider a generalization of the Busemann-Petty problem (the BPGM problem): given two convex origin-symmetric bodies K and L in $\mathbb{R}^n$ such that, for every $\xi \in S^{n-1}$,

$$\mu(K \cap \xi^\perp) \le \mu(L \cap \xi^\perp),$$

does it follow that

$$\mu(K) \le \mu(L)?$$

Since f is continuous and strictly positive, the answer is affirmative if $n = 2$. We start with a formula expressing the measure of sections in terms of the Fourier transform, generalizing the result of Theorem 3.8.

THEOREM 5.13. *Let K be an origin-symmetric star body in $\mathbb{R}^n$, then, for every $\xi \in S^{n-1}$,*

$$\mu(K \cap \xi^\perp) = \frac{1}{\pi} \left(|x|_2^{-n+1} \int_0^{|x|_2/\|x\|_K} t^{n-2} f\left(\frac{tx}{|x|_2}\right) dt \right)^\wedge (\xi),$$

where the Fourier transform of the function of $x \in \mathbb{R}^n$ in the right-hand side is a continuous homogeneous of degree -1 function on $\mathbb{R}^n \setminus \{0\}$.

PROOF. If χ is the indicator function of the interval $[-1, 1]$, then, passing to the polar coordinates in the hyperplane $\xi^\perp$, we get

$$\mu(K \cap \xi^\perp) = \int_{(x,\xi)=0} \chi(\|x\|_K) f(x) dx$$

$$= \int_{S^{n-1}\cap\xi^\perp} \left(\int_0^{\|\theta\|_K^{-1}} t^{n-2} f(t\theta) dt \right) d\theta.$$

In order to apply Lemma 3.7, we extend the function under the outer integral to a homogeneous of degree $-n+1$ continuous function on $\mathbb{R}^n \setminus \{0\}$ (note that $|x|_2 = 1$, so we do not change the value of the integral). By Lemma 3.7, the spherical Radon transform of such a function is equal (up to a constant) to the restriction to the sphere of its Fourier transform, so

$$\mu(K \cap \xi^\perp) = \int_{S^{n-1}\cap\xi^\perp} |x|_2^{-n+1} \left(\int_0^{\frac{|x|_2}{\|x\|_K}} t^{n-2} f\left(\frac{tx}{|x|_2}\right) dt \right) dx \tag{5.17}$$

$$= R \left(|x|_2^{-n+1} \int_0^{\frac{|x|_2}{\|x\|_K}} t^{n-2} f\left(\frac{tx}{|x|_2}\right) dt \right) (\xi)$$

$$= \frac{1}{\pi} \left(|x|_2^{-n+1} \int_0^{\frac{|x|_2}{\|x\|_K}} t^{n-2} f\left(\frac{tx}{|x|_2}\right) dt \right)^\wedge (\xi).$$

$\square$

The following elementary observation will play the same role as Hölder's inequality in the solution of the original Busemann-Petty problem in Section 5.1.

LEMMA 5.14. *Let $a, b > 0$ and let α be a non-negative function on $(0, \max\{a, b\}]$ so that the integrals below converge. Then*

$$\int_0^a t^{n-1}\alpha(t)dt - a\int_0^a t^{n-2}\alpha(t)dt \le \int_0^b t^{n-1}\alpha(t)dt - a\int_0^b t^{n-2}\alpha(t)dt. \tag{5.18}$$

PROOF. Inequality (5.18) is equivalent to

$$a\int_a^b t^{n-2}\alpha(t)dt \le \int_a^b t^{n-1}\alpha(t)dt.$$

Note that the latter inequality also holds in the case $a \ge b$. □

The next theorem generalizes the first Lutwak connection.

THEOREM 5.15. *(BPGM: affirmative case) Let f and μ be as above. Consider an origin-symmetric star body K in $\mathbb{R}^n$ such that $\|\cdot\|_K^{-1}$ is a positive definite distribution on $\mathbb{R}^n$. Then for any origin-symmetric star body L in $\mathbb{R}^n$ satisfying*

$$\mu(K \cap \xi^\perp) \le \mu(L \cap \xi^\perp), \ \forall \xi \in S^{n-1}, \tag{5.19}$$

we have

$$\mu(K) \le \mu(L).$$

PROOF. By Theorem 5.13, condition (5.19) is equivalent to an inequality for two continuous functions on the sphere:

$$\left(|x|_2^{-n+1} \int_0^{\frac{|x|_2}{\|x\|_K}} t^{n-2} f\left(\frac{tx}{|x|_2}\right) dt \right)^\wedge (\xi)$$

$$\le \left(|x|_2^{-n+1} \int_0^{\frac{|x|_2}{\|x\|_L}} t^{n-2} f\left(\frac{tx}{|x|_2}\right) dt \right)^\wedge (\xi)$$

for each $\xi \in S^{n-1}$. We integrate the latter inequality over S^{n-1} with respect to the measure μ_0 corresponding to the positive definite homogeneous of degree -1 distribution $\|\cdot\|_K^{-1}$ by Corollary 2.26:

$$\int_{S^{n-1}} \left(|x|_2^{-n+1} \int_0^{\frac{|x|_2}{\|x\|_K}} t^{n-2} f\left(\frac{tx}{|x|_2}\right) dt \right)^\wedge (\xi)\, d\mu_0(\xi)$$

$$\le \int_{S^{n-1}} \left(|x|_2^{-n+1} \int_0^{\frac{|x|_2}{\|x\|_L}} t^{n-2} f\left(\frac{tx}{|x|_2}\right) dt \right)^\wedge (\xi)\, d\mu_0(\xi),$$

and now apply the spherical Parseval formula, Corollary 3.23:

$$\int_{S^{n-1}} \|x\|_K^{-1} \left(\int_0^{\|x\|_K^{-1}} t^{n-2} f(tx)dt \right) dx$$

$$\leq \int_{S^{n-1}} \|x\|_K^{-1} \left(\int_0^{\|x\|_L^{-1}} t^{n-2} f(tx)dt \right) dx. \tag{5.20}$$

By Lemma 5.14 with $a = \|x\|_K^{-1}$, $b = \|x\|_L^{-1}$ and $\alpha(t) = f(tx)$,

$$\int_0^{\|x\|_K^{-1}} t^{n-1} f(tx)\, dt - \|x\|_K^{-1} \int_0^{\|x\|_K^{-1}} t^{n-2} f(tx)\, dt$$

$$\leq \int_0^{\|x\|_L^{-1}} t^{n-1} f(tx)dt - \|x\|_K^{-1} \int_0^{\|x\|_L^{-1}} t^{n-2} f(tx)dt, \quad \forall x \in S^{n-1}.$$

Integrating over S^{n-1}, we get

$$\int_{S^{n-1}} \left(\int_0^{\|x\|_K^{-1}} t^{n-1} f(tx)dt \right) dx - \int_{S^{n-1}} \|x\|_K^{-1} \left(\int_0^{\|x\|_K^{-1}} t^{n-2} f(tx)dt \right) dx \tag{5.21}$$

$$\leq \int_{S^{n-1}} \left(\int_0^{\|x\|_L^{-1}} t^{n-1} f(tx)dt \right) dx - \int_{S^{n-1}} \|x\|_K^{-1} \left(\int_0^{\|x\|_L^{-1}} t^{n-2} f(tx)dt \right) dx.$$

Adding equations (5.20) and (5.21), we get

$$\int_{S^{n-1}} \left(\int_0^{\|x\|_K^{-1}} t^{n-1} f(tx)dt \right) dx \leq \int_{S^{n-1}} \left(\int_0^{\|x\|_L^{-1}} t^{n-1} f(tx)dt \right) dx.$$

It remains to note that the left- and right-hand sides of the latter inequality are equal to $\mu(K)$ and $\mu(L)$, respectively, written in polar coordinates. □

We need a slightly modified version of the perturbation argument used in the proof of Theorem 5.3.

LEMMA 5.16. *Consider an infinitely smooth origin-symmetric convex body L with positive curvature and even functions $f, g \in C^2(\mathbb{R}^n)$, such that f is strictly positive on $\mathbb{R}^n$. For $\varepsilon > 0$ define a star body D by an equation for its radial function $\|x\|_D^{-1}$:*

$$\int_0^{\|x\|_D^{-1}} t^{n-2} f(tx)dt = \int_0^{\|x\|_L^{-1}} t^{n-2} f(tx)dt - \varepsilon g(x), \quad \forall x \in S^{n-1}.$$

Then if ε is small enough, the body D is convex.

PROOF. For small enough ε, define a function $\alpha_\varepsilon(x)$ on S^{n-1} so that

$$\int_0^{||x||_L^{-1}} t^{n-2} f(tx)dt - \varepsilon g(x) = \int_0^{||x||_L^{-1}-\alpha_\varepsilon(x)} t^{n-2} f(tx)dt, \ \forall x \in S^{n-1}. \tag{5.22}$$

Using monotonicity of $\int_0^y t^{n-2} f(tx)dt$, for $y \in \mathbb{R}^+$ (recall that $f(tx) > 0$, for $tx \in \mathbb{R}^n \setminus \{0\}$), we get

$$||x||_D^{-1} = ||x||_L^{-1} - \alpha_\varepsilon(x), \ \forall x \in S^{n-1}. \tag{5.23}$$

Moreover, using that f and its partial derivatives of the order one are bounded for finite values of x, we get, from (5.22), that $\alpha_\varepsilon(x)$ and its first and second derivatives converge uniformly to 0 (for $x \in S^{n-1}$, as $\varepsilon \to 0$). Using that L is convex with positive curvature, one can choose a small enough ε so that the body D is convex (with positive curvature). □

We now generalize the second Lutwak connection.

THEOREM 5.17. *(BPGM: negative case) Let f be an even continuous positive function so that $f \in C^\infty(\mathbb{R}^n)$. If L is an infinitely smooth origin-symmetric convex body in $\mathbb{R}^n$ with positive curvature and the function $\|\cdot\|_L^{-1}$ does not represent a positive definite distribution, then there exists a convex origin-symmetric body D in $\mathbb{R}^n$ so that*

$$\mu(D \cap \xi^\perp) \le \mu(L \cap \xi^\perp), \ \forall \xi \in S^{n-1},$$

but

$$\mu(D) > \mu(L).$$

PROOF. Since L is infinitely smooth, by Corollary 3.17, $(\|\cdot\|_L^{-1})^\wedge$ is a continuous function on S^{n-1}. This function must be negative on some open symmetric subset Ω of S^{n-1}. Consider a non-negative even function $h \in C^\infty(S^{n-1})$ supported in Ω. Extend h to a homogeneous function $h(\theta)r^{-1}$ of degree -1 on the whole $\mathbb{R}^n$. Then, by Lemma 3.16, the Fourier transform of $h(\theta)r^{-1}$ is a function $g(\theta)r^{-n+1}$, where $g \in C^\infty(S^{n-1})$.

For $\varepsilon > 0$, we define a body D by

$$|x|_2^{-n+1} \int_0^{\frac{|x|_2}{\|x\|_D}} t^{n-2} f\left(\frac{tx}{|x|_2}\right) dt$$

$$= |x|_2^{-n+1} \int_0^{\frac{|x|_2}{\|x\|_L}} t^{n-2} f\left(\frac{tx}{|x|_2}\right) dt - \varepsilon g\left(\frac{x}{|x|_2}\right) |x|_2^{-n+1}.$$

By Lemma 5.16, one can choose a small enough ϵ so that the body D is convex.

Since $h \ge 0$, by Theorem 5.13, we have

$$\mu(D \cap \xi^\perp) = \frac{1}{\pi} \left(|x|_2^{-n+1} \int_0^{|x|_2/||x||_D} t^{n-2} f\left(\frac{tx}{|x|_2}\right) dt \right)^\wedge (\xi)$$

$$= \frac{1}{\pi}\left(|x|_2^{-n+1}\int_0^{|x|_2/\|x\|_L} t^{n-2} f\left(\frac{tx}{|x|_2}\right) dt\right)^{\wedge}(\xi) - \frac{(2\pi)^n \varepsilon h(\xi)}{\pi}$$

$$\leq \mu(L\cap\xi^\perp).$$

On the other hand, the function h is positive only where $(\|\cdot\|_L^{-1})^\wedge(\xi)$ is negative so, for every $\xi \in S^{n-1}$,

$$(\|\cdot\|_L^{-1})^\wedge(\xi)\left(|x|_2^{-n+1}\int_0^{|x|_2/\|x\|_D} t^{n-2} f\left(\frac{tx}{|x|_2}\right) dt\right)^{\wedge}(\xi)$$

$$= (\|\cdot\|_L^{-1})^\wedge(\xi)\left(|x|_2^{-n+1}\int_0^{|x|_2/\|x\|_L} t^{n-2} f\left(\frac{tx}{|x|_2}\right) dt\right)^{\wedge}(\xi)$$

$$-(2\pi)^n(\|\cdot\|_L^{-1})^\wedge(\xi)\ \varepsilon h(\xi)$$

$$\geq (\|\cdot\|_L^{-1})^\wedge(\xi)\left(|x|_2^{-n+1}\int_0^{|x|_2/\|x\|_L} t^{n-2} f\left(\frac{tx}{|x|_2}\right) dt\right)^{\wedge}(\xi),$$

with strict inequality for ξ from some open set.

Integrate the latter inequality over S^{n-1} and apply the spherical Parseval identity, Lemma 3.22. The same computations (based on Lemma 5.14) as in the proof of Theorem 5.15 prove that $\mu_n(D) > \mu_n(L)$. □

THEOREM 5.18. *If f is an even continuous positive function on $\mathbb{R}^n$, then the answer to the BPGM problem is affirmative for $n \leq 4$. If f also belongs to $C^\infty(\mathbb{R}^n)$, then the answer to the problem is negative for each $n \geq 5$.*

PROOF. First, we use the result of Corollary 4.9 that for any origin-symmetric convex body K in $\mathbb{R}^n$, $n \leq 4$, $\|\cdot\|_K^{-1}$ represents a positive definite distribution. The affirmative answer to BPGM for $n \leq 4$ follows immediately from Theorem 5.15.

Let $n \geq 5$. By Theorem 4.13 (with $k = 1$), for any $q > 2$ the unit ball of the space ℓ_q^n is not an intersection body. By Lemma 4.10 there exists an infinitely smooth origin-symmetric convex body $L \subset \mathbb{R}^n$ with positive curvature which is not an intersection body, so $\|\cdot\|_L^{-1}$ is not a positive definite distribution. The result follows from Theorem 5.17. □

The result of Theorem 5.18 holds for functions f satisfying weaker conditions, where the proof has to be slightly adjusted for a given function f; see [Zv2], [Y1].

5.5. Remarks and further results

The history of the solution of the Busemann-Petty problem is presented in the Introduction. After the solution was completed in 1997, several papers appeared where the authors simplified or generalized the solution. Barthe, Fradelizi and Maurey [BFM] replaced the results of Theorems 3.18 and 4.1 in the solution by a geometric argument, which directly relates intersection bodies to the derivatives of parallel section functions. This approach allows us to avoid the Fourier transform and analytic continuation. A proof of the affirmative part using spherical harmonics, which we presented in Section 5.3, was found in [K16]. The generalization of the Busemann-Petty problem from Section 5.2 was proved in [K13]. The result of Zvavitch was first proved for Gaussian measures in [Zv1]. The general result, presented in Section 5.4, appears in [Zv2] in even more general form. It is actually possible to get some results when different measures are used for the bodies and for sections. This leads to several new results of Busemann-Petty type. For example, the following theorem was proved in [Zv2]:

THEOREM 5.19. *Let K and L be origin-symmetric convex bodies in $\mathbb{R}^n$ and let $0 < p \leq 2$ so that for every $\xi \in S^{n-1}$,*

$$\int_{K\cap\xi^\perp} \|x\|_p dx \leq \int_{L\cap\xi^\perp} \|x\|_p dx.$$

Then $\mathrm{Vol}_n(K) \leq \mathrm{Vol}_n(L)$.

The latter statement is not true if $p > 2$.

It is natural to ask the question of the Busemann-Petty problem for sections of lower dimensions. Let D and L be origin-symmetric convex bodies in $\mathbb{R}^n$, $1 \leq m \leq n-1$. Suppose that for every m-dimensional subspace H of $\mathbb{R}^n$,

$$\mathrm{Vol}_m(D\cap H) \leq \mathrm{Vol}_m(L\cap H). \tag{5.24}$$

Does it follow that

$$\mathrm{Vol}_n(D) \leq \mathrm{Vol}_n(L)?$$

A connection between this lower dimensional Busemann-Petty problem and generalized k-intersection bodies (see Section 4.5) was found by Zhang [Zh3] (see also [RZ]) who proved that the answer to the lower dimensional Busemann-Petty problem with $m = n-k$ is affirmative if and only if every origin-symmetric convex body in $\mathbb{R}^n$ is a generalized k-intersection body. Let us show one direction of this result.

THEOREM 5.20. *Suppose that $1 \leq k < n$ and that D is an origin-symmetric star body in $\mathbb{R}^n$ that is a generalized k-intersection body; see Section 4.5 for the definition. Then for any origin-symmetric star body L in $\mathbb{R}^n$ such that*

$$\mathrm{Vol}_{n-k}(D\cap H) \leq \mathrm{Vol}_{n-k}(L\cap H) \qquad \forall H \in G(n, n-k), \tag{5.25}$$

we have $\mathrm{Vol}_n(D) \leq \mathrm{Vol}_n(L)$.

PROOF. By (2.3), condition (5.25) is equivalent to

$$R_{n-k}(\|\cdot\|_D^{-n+k})(H) \leq R_{n-k}(\|\cdot\|_L^{-n+k})(H) \tag{5.26}$$

for every $H \in G(n, n-k)$, where R_{n-k} is the $(n-k)$-dimensional spherical Radon transform defined as an operator from $C(S^{n-1})$ to $C(G(n, n-k))$ such that

$$R_{n-k}f(H) = \int_{S^{n-1}\cap H} f(x)\, dx.$$

The definition of a generalized k-intersection body implies that there exists a positive linear functional μ on $X = R_{n-k}(C(S^{n-1}))$ so that

$$\int_{S^{n-1}} \|x\|_D^{-k} f(x)\, dx = \mu(R_{n-k}f)$$

for every $f \in C(S^{n-1})$. Put $f_1(\cdot) = \|\cdot\|_D^{-n+k}$, $f_2(\cdot) = \|\cdot\|_L^{-n+k}$. By (5.26) and the positivity of μ, we have

$$\mu(R_{n-k}f_1) \leq \mu(R_{n-k}f_2),$$

so

$$\int_{S^{n-1}} \|x\|_D^{-k} \|x\|_D^{-n+k}\, dx \le \int_{S^{n-1}} \|x\|_D^{-k} \|x\|_L^{-n+k}\, dx,$$

and we can finish the proof by using Hölder's inequality in the same way as in Theorem 5.6. □

Using the connection between the lower dimensional Busemann-Petty problem and generalized k-intersection bodies, Bourgain and Zhang [BZ] proved that the answer to the lower dimensional problem is negative if the dimension of sections $m > 3$. Another proof of the same fact was given in [K14]. In fact, by Theorem 4.23, every generalized k-intersection body is also a k-intersection body. But if $k < n-3$, then the unit balls of the spaces ℓ_q^n, $q > 2$, are not k-intersection bodies (Theorem 4.13). This means that if $m = n-k > 3$, there exist origin-symmetric convex bodies in $\mathbb{R}^n$ that are not generalized k-intersection bodies. By Zhang's connection, the answer to the lower dimensional Busemann-Petty problem is negative in this case.

The lower dimensional Busemann-Petty problem is still open for $m = 2, 3$. Bourgain and Zhang [BZ] also proved that if $m = 2$, then it is not possible to construct a counterexample by small perturbations of the Euclidean ball. A related question is whether it is true that every k-intersection body is a generalized k-intersection body. If the answer to the latter question were affirmative, then the answer to the lower dimensional Busemann-Petty problem would also be affirmative for $m = 2, 3$. This follows from the fact that every origin-symmetric convex body in $\mathbb{R}^n$ is an $(n-3)$- and $(n-2)$-intersection body, Corollary 4.9.

V. Yaskin [Y1] has formulated and solved the Busemann-Petty problem in hyperbolic and spherical spaces. It appears that the answer in the spherical case is the same as for the original problem, while in the hyperbolic case the answer is affirmative only in dimension 2. The proof uses Zvavitch's result with different measures for the bodies and their sections. In another very recent paper [Y2], Yaskin solved the lower dimensional Busemann-Petty problem in the hyperbolic space. The answer is negative for sections of all dimensions $m \ge 2$.

For more generalizations of the Busemann-Petty problem and other results of Busemann-Petty type, see [K14], [K15], [K17], [KYY], [RZ], [Rb2].

The isomorphic version of the Busemann-Petty problem is open and is equivalent to the famous slicing problem; see [MiP]. Namely, it is not known whether there exists an absolute constant c so that, for any origin-symmetric convex bodies K, L in $\mathbb{R}^n$ satisfying

$$\mathrm{Vol}_{n-1}(K \cap H) \le \mathrm{Vol}_{n-1}(L \cap H)$$

for all central hyperplanes H, we necessarily have

$$\mathrm{Vol}_n(K) \le c\mathrm{Vol}_n(L).$$

Recall that the slicing problem asks whether there exists an absolute constant c so that for any origin-symmetric convex body K with volume 1 there exists a hyperplane section whose $(n-1)$-dimensional volume is greater than c. The best known lower estimate belongs to Bourgain [Bo2] and depends on the dimension (up to a logarithmic term) as $n^{-1/4}$. We refer the reader to the paper by Klartag [Kla] for related references and the most recent status of the problem.

Finally, let us note that the second Lutwak connection, Theorem 5.2, is presented here in the form of [Ga1] and [Zh1], [Zh2], which slightly improves the original result of Lutwak [Lu1]. The generalized Lutwak connections from Section 5.2 were proved in [K13].

CHAPTER 6

Intersection Bodies and L_p-Spaces

6.1. L_p-spaces and positive definite functions

Let K be an origin-symmetric star body in $\mathbb{R}^n$. We say that a Borel probability measure μ on $\mathbb{R}^n$ is *K-isotropic* if, for all $x \in \mathbb{R}^n$, $x \neq 0$, the functions

$$\xi \mapsto \frac{(x,\xi)}{\|x\|_K}, \qquad \xi \in \mathbb{R}^n,$$

are identically distributed with respect to μ.

Geometrically, this means that all one-dimensional projections of the measure μ are equal up to a scalar parameter, namely, the measures ν_x on $\mathbb{R}$ defined by

$$\nu_x([a,b]) = \mu\{\xi \in \mathbb{R}^n : \frac{(x,\xi)}{\|x\|_K} \in [a,b]\}, \quad a,b \in \mathbb{R}, \tag{6.1}$$

are equal for all choices of $x \in \mathbb{R}^n$, $x \neq 0$.

The probabilistic meaning of the concept of a K-isotropic measure is as follows: if μ is a K-isotropic measure on $\mathbb{R}^n$ and a random vector $X = (X_1, \dots, X_n)$ has the probability distribution μ, then linear combinations $\sum_{k=1}^n \alpha_k X_k$ are identically distributed for all choices of $\alpha = (\alpha_1, \dots, \alpha_n) \in \mathbb{R}^n$ with $\|\alpha\|_K = 1$.

LEMMA 6.1. *A finite Borel measure μ on $\mathbb{R}^n$ is K-isotropic if and only if $\widehat{\mu} = f(\|\cdot\|_K)$, where f is an even continuous function on $\mathbb{R}$.*

PROOF. Consider an arbitrary $x \in \mathbb{R}^n$, $x \neq 0$, and let ν_x be the measure on $\mathbb{R}$ defined by (6.1). Then, for every $u \in \mathbb{R}$,

$$\begin{aligned}\widehat{\mu}(ux) &= \int_{\mathbb{R}^n} e^{-i(ux,\xi)}\, d\mu(\xi)\\ &= \int_{\mathbb{R}^n} e^{-iu\|x\|_K \frac{(x,\xi)}{\|x\|_K}}\, d\mu(\xi)\\ &= \int_{\mathbb{R}} e^{-iuy\|x\|_K}\, d\nu_x(y) = \widehat{\nu_x}(u\|x\|_K).\end{aligned}$$

If μ is an isotropic measure, the measure ν_x is independent of the choice of x. We set $f = \widehat{\nu_x}$. Then $\widehat{\mu}(x) = f(\|x\|_K)$ for every $x \in \mathbb{R}^n$.

Conversely, if $\widehat{\mu} = f(\|\cdot\|_K)$ for some function f, then, for each $x \in \mathbb{R}^n$, $x \neq 0$, we have $f(|u|\|x\|_K) = \widehat{\nu_x}(u\|x\|_k)$ for every $u \in \mathbb{R}$, so $f = \widehat{\nu_x}$. Therefore, all the measures ν_x are equal, and μ is K-isotropic. □

A complex-valued function f defined on $\mathbb{R}^n$ is called *positive definite* on $\mathbb{R}^n$, if, for every finite sequence $\{x_i\}_{i=1}^m$ in $\mathbb{R}^n$ and every choice of complex numbers

$\{c_i\}_{i=1}^m$, we have

$$\sum_{i=1}^m \sum_{j=1}^m c_i \bar{c}_j f(x_i - x_j) \geq 0.$$

An immediate consequence of this definition is that the pointwise limit of positive definite functions is also a positive definite function.

The classical theorem of Bochner (see [F, Ch. 19, Section 2] for a proof) states that

THEOREM 6.2. *(Bochner's Theorem) A continuous function on $\mathbb{R}^n$ is positive definite if and only if it is the Fourier transforms of a finite Borel measure on $\mathbb{R}^n$.*

Bochner's theorem implies that the product of positive definite functions is a positive definite function. In fact, if f_1 and f_2 are positive definite functions on $\mathbb{R}^n$, and μ_1 and μ_2 are finite measures on $\mathbb{R}^n$ whose Fourier transforms are equal to f_1 and f_2, respectively, then the product $f_1 f_2$ is the Fourier transform of the convolution $\mu_1 * \mu_2$, which is a finite Borel measure on $\mathbb{R}^n$ defined by

$$\mu_1 * \mu_2(A) = \int_{\mathbb{R}^n} \mu_1(A - x)\, d\mu_2(x)$$

for every Borel subset A of $\mathbb{R}^n$. Again by Bochner's theorem, the product $f_1 f_2$ is a positive definite function on $\mathbb{R}^n$.

It follows from Bochner's theorem and Lemma 6.1 that to find all K-isotropic measures on $\mathbb{R}^n$, it suffices to characterize the class of all positive definite functions of the form $f(\|\cdot\|_K)$.

The classical example of an isotropic measure is the standard Gaussian measure on $\mathbb{R}^n$. This measure is isotropic with respect to the unit Euclidean ball. Another well-known example, which we use in the future, is that of q-stable measures. Let $0 < q \leq 2$. As follows from Lemma 2.27, the function $t \mapsto \exp(-|t|^q)$ is positive definite on $\mathbb{R}$, and, hence, the function $g = \exp(-\|\cdot\|_q^q)$ is positive definite on $\mathbb{R}^n$, as the product of positive definite functions. By Bochner's theorem, there exists a finite Borel measure μ_q on $\mathbb{R}^n$ whose Fourier transform is equal to g. The measure μ_q is called the *standard q-stable measure* on $\mathbb{R}^n$. By Lemma 6.1, this measure is B_q^n-isotropic. Therefore, if $\eta_1, ..., \eta_n$ are independent identically distributed random variables, whose joint distribution is μ_q (these random variables are called symmetric q-stable; see Section 2.8), then linear combinations of these random variables are equally distributed provided that the ℓ_q^n-norm of the coefficient vector is 1:

LEMMA 6.3. *Let $\eta_1, ..., \eta_n$ be independent normalized symmetric q-stable random variables. Then, for any $a_1, ..., a_n \in \mathbb{R}^n$, the random variables*

$$\sum_{k=1}^n a_k \eta_k \quad \text{and} \quad \left(\sum_{k=1}^n |a_k|^q \right)^{1/q} \eta_1$$

have equal distributions.

Positive definite norm dependent functions are related to isometric embeddings of normed spaces in L_p. We say that a space $(\mathbb{R}^n, \|\cdot\|_K)$ embeds isometrically in L_p, $p > 0$, if there exists a linear operator $T : \mathbb{R}^n \mapsto L_p([0,1])$ so that for every $x \in \mathbb{R}^n$, $\|x\|_K = \|Tx\|_{L_p}$. A well-known fact going back to P. Levy is that the norms of subspaces of L_p admit a special representation (6.2), which we call the *Levy representation.*

LEMMA 6.4. *Let K be an origin-symmetric star body in $\mathbb{R}^n$, $p > 0$. The space $(\mathbb{R}^n, \|\cdot\|_K)$ embeds isometrically in L_p if and only if there exists a finite Borel measure μ on the sphere S^{n-1} so that, for every $x \in \mathbb{R}^n$,*

$$\|x\|_K^p = \int_{S^{n-1}} |(x,\xi)|^p d\mu(\xi). \tag{6.2}$$

PROOF. Suppose that $(\mathbb{R}^n, \|\cdot\|_K)$ embeds isometrically in L_p and that T is the corresponding linear operator. Denote $f_i = Te_i$, where $\{e_i\}_{i=1}^n$ is the standard basis in $\mathbb{R}^n$. Let ν be the *joint distribution* of the functions $f_1, ... f_n$, i.e. ν is a measure on $\mathbb{R}^n$ defined by

$$\nu(B) = \lambda\{t \in [0,1] : (f_1(t), ..., f_n(t)) \in B\}$$

for every Borel set $B \subset \mathbb{R}^n$, where λ is Lebesgue measure on $[0, 1]$.

Define the p-projection of ν to the sphere S^{n-1} by

$$\mu(A) = \int_{A\times\mathbb{R}} |x|_2^p \, d\nu(x)$$

for every Borel set $A \subset S^{n-1}$. By the definition of the measure ν,

$$\int_{\mathbb{R}^n} |x_i|^p \, d\nu(x) = \int_0^1 |f_i|^p \, d\lambda < \infty,$$

for every $i = 1, ..., n$. It follows that μ is a finite measure on S^{n-1}, because

$$|x|_2^p \le C \sum_{i=1}^n |x_i|^p$$

for some constant C.

For every $x \in \mathbb{R}^n$, we have

$$\|x\|_K^p = \|\sum_{i=1}^n x_i Te_i\|_{L_p}^p = \int_0^1 |\sum_{i=1}^n x_i f_i(t)|^p \, d\lambda(t)$$

$$= \int_{\mathbb{R}^n} |(x,y)|^p \, d\nu(y) = \int_{\mathbb{R}^n} |(x, \frac{y}{|y|_2})|^p |y|_2^p \, d\nu(y).$$

After the change of variables $\xi = y/|y|_2$ the latter integral becomes

$$\int_{S^{n-1}} |(x,\xi)|^p \, d\mu(\xi).$$

Conversely, given (6.2), define an operator T from $(\mathbb{R}^n, \|\cdot\|_K)$ into $L_p(S^{n-1}, \mu)$ by $Tx = (x, \cdot) \in L_p(S^{n-1}, \mu)$ for every $x \in R^n$. Clearly, this operator is a linear isometry. It is well known (see for example [LT, p. 124]) that every separable L_p-space embeds isometrically in $L_p([0,1])$, which finishes the proof. $\square$

The next two results establish a connection between embeddings in L_p and positive definite functions.

LEMMA 6.5. *Let K be an origin-symmetric star body in $\mathbb{R}^n$, $p > 0$, and suppose there exists a non-constant even function f on $\mathbb{R}$ so that $f(\|\cdot\|_K)$ is a positive definite function. By Bochner's theorem, Theorem 6.2, there exists a measure ν on*

$\mathbb{R}$ satisfying $\hat{\nu} = f$. Suppose, in addition, that ν has a finite moment of the order p, i.e.

$$\int_{\mathbb{R}} |t|^p d\nu(t) < \infty.$$

Then the space $(\mathbb{R}^n, \|\cdot\|_K)$ embeds isometrically in L_p.

PROOF. The function $f(\|\cdot\|_K)$ is positive definite, so by Bochner's theorem, Theorem 6.2, there exists a finite measure μ on R^n so that $\hat{\mu} = f(\|\cdot\|_K)$. By Lemma 6.1, the measure μ is K-isotropic, and, for every $x \in \mathbb{R}^n$, we have

$$\int_{\mathbb{R}^n} |(x,\xi)|^p \, d\mu(\xi) = \|x\|_K^p \int_{\mathbb{R}^n} \left|\frac{(x,\xi)}{\|x\|_K}\right|^p \, d\mu(\xi)$$

$$= \|x\|_K^p \int_{\mathbb{R}} |t|^p \, d\nu(t).$$

The result follows from Lemma 6.4. □

THEOREM 6.6. *Let K be an origin-symmetric star body in $\mathbb{R}^n$ and let $0 < p \le 2$. Then the space $(\mathbb{R}^n, \|\cdot\|_K)$ embeds isometrically in L_p if and only if the function $\exp(-\|\cdot\|_K^p)$ is positive definite.*

PROOF. The necessity follows from an argument similar to that in the beginning of Section 3.1. Consider any even positive definite function f on $\mathbb{R}$, and fix $\xi \in S^{n-1}$. Then the function $x \mapsto f((x,\xi))$ is positive definite on $\mathbb{R}^n$. In fact, if σ is the measure on $\mathbb{R}$, whose Fourier transform is f (such measure exists by Bochner's theorem, Theorem 6.2), then a simple calculation shows that the function $f((x,\xi))$ is the Fourier transform of a measure σ_ξ in $\mathbb{R}^n$, which is the measure σ supported in the line $t\xi$, $t \in \mathbb{R}$:

$$\int_{\mathbb{R}^n} e^{-i(x,\eta)} d\sigma_\xi(\eta) = \int_{\mathbb{R}} e^{-it(x,\xi)} d\sigma(t)$$

$$= \hat{\sigma}((x,\xi)) = f((x,\xi)).$$

By Bochner's theorem, Theorem 6.2, the function $f((x,\xi))$ is positive definite.

By Lemma 2.27, for $0 < p \le 2$, the function $t \mapsto \exp(-|t|^p)$ is positive definite on $\mathbb{R}$. By the argument above, for every fixed $\xi \in S^{n-1}$, the function $x \mapsto \exp(-|(x,\xi)|^p)$ is positive definite on $\mathbb{R}^n$.

Suppose that $(\mathbb{R}^n, \|\cdot\|_K)$ embeds in L_p. Then, writing the integral in the representation (6.2) as the limit of its Riemann sums, we express the function $\exp(-\|\cdot\|_K^p)$ as the limit of products of positive definite functions of the form $\exp(-c_i|(x,\xi_i)|^p)$. The result follows from the fact that positive definiteness is preserved under limits and products; see above.

Conversely, suppose that the function $\exp(-\|\cdot\|_K^p)$ is positive definite. The Fourier transform of the function $f(t) = \exp(-|t|^p)$ is the function γ_p. This function has finite moments of all orders $r < p$ (see Lemma 2.29). By Lemma 6.5, the space $(\mathbb{R}^n, \|\cdot\|_K)$ embeds in L_r for every $r < p$. By Lemma 6.4, there exist finite measures μ_r on S^{n-1} so that, for every $x \in \mathbb{R}^n$,

$$\|x\|_K^r = \int_{S^{n-1}} |(x,\xi)|^r d\mu_r(\xi). \tag{6.3}$$

Since $r < 2$ and $|\xi|_2 = 1$, we have for every $\xi \in S^{n-1}$,

$$1 = \sum_{i=1}^{n} (e_i, \xi)^2 \le \sum_{i=1}^{n} |(e_i, y)|^r,$$

where $\{e_i\}_{i=1}^n$ is the standard basis in $\mathbb{R}^n$. Hence,

$$\mu_r(S^{n-1}) = \int_{S^{n-1}} d\mu_r(\xi) \le \sum_{i=1}^{n} \int_{S^{n-1}} |(e_i, \xi)|^r d\mu_r(\xi)$$

$$= \sum_{i=1}^{n} \|e_i\|_K^r < C = \max_{r\in[0,2]} \sum_{i=1}^{n} \|e_i\|_K^r,$$

so the constant C depends on n but not on r. We see that the measures μ_r are uniformly bounded. By Alaoglu's theorem, [DS, p. 424], the unit ball of the dual space $(C(S^{n-1}))^*$ is compact in the w^*-topology, so we can find a sequence $r_j \to p$ so that the measures μ_{r_j} converge weakly to some finite measure μ on S^{n-1}, in the sense that, for any $f \in C(S^{n-1})$,

$$\int_{S^{n-1}} f(\xi)\, d\mu_{r_j}(\xi) \to \int_{S^{n-1}} f(\xi)\, d\mu(\xi).$$

By a simple limiting argument, the right-hand side of (6.3) converges to

$$\int_{S^{n-1}} (|(x, \xi)|^{r_j} - |(x, \xi)|^p) d\mu_{r_j}(\xi) + \int_{S^{n-1}} |(x, \xi)|^p d\mu_{r_j}(\xi)$$

$$\to \int_{S^{n-1}} |(x, \xi)|^p d\mu(\xi),$$

so the measure μ satisfies

$$\|x\|_K^p = \int_{S^{n-1}} |(x, \xi)|^p d\mu(\xi),$$

and by Lemma 6.4, the space $(\mathbb{R}^n, \|\cdot\|_K)$ embeds in L_p. □

Let us show a couple of applications that will be important for us later.

COROLLARY 6.7. *Let $0 < p < q \le 2$, and let K be an origin-symmetric star body in $\mathbb{R}^n$. If the space $(\mathbb{R}^n, \|\cdot\|_K)$ embeds isometrically in L_q, then it also embeds in L_p.*

PROOF. We use equality (2.49): for every $\alpha \in (0, 1)$, there exists a finite measure μ_α on $[0, \infty)$ so that for each $z \ge 0$,

$$\exp(-z^\alpha) = \int_0^\infty \exp(-tz)\, d\mu_\alpha(t).$$

Put $z = \|x\|_K^q$ and $\alpha = p/q < 1$. We get that for every $x \in \mathbb{R}^n$,

$$\exp(-\|x\|^p) = \int_0^\infty \exp(-t\|x\|_K^q) d\mu_{p/q}(t).$$

Since $(\mathbb{R}^n, \|\cdot\|_K)$ embeds in L_q, by Theorem 6.6, the functions $x \mapsto \exp(-t\|x\|_K^q)$ are positive definite on $\mathbb{R}^n$ for all $t > 0$. Clearly, the function $\exp(-\|\cdot\|_K^p)$ is also positive definite (apply the definition of a positive definite function directly), which, again by Theorem 6.6, means that $(\mathbb{R}, \|\cdot\|_K)$ embeds in L_p. □

COROLLARY 6.8. *Let K be an origin-symmetric convex body in $\mathbb{R}^2$. Then the normed space $(\mathbb{R}^n, \|\cdot\|_K)$ embeds isometrically in L_1.*

PROOF. By the Cauchy formula (2.14), for every $\theta \in [0, 2\pi]$, the length of the projection of K to the hyperplane $\theta^\perp$ is equal to

$$\mathrm{Vol}_1\left(K\middle|\theta^\perp\right) = \frac{1}{2}\int_0^{2\pi} |(\theta, v)| dS(K, v),$$

where $S(K, \cdot)$ is the surface area measure of K. The crucial observation is that in dimension 2 the length of the projection is equal to twice the support functional (dual norm) of K in the direction $\theta + \pi/2$. The support function of K is the norm of the polar body K^*, so, by Lemma 6.4, the space $(\mathbb{R}^2, \|\cdot\|_{K^*})$ embeds in L_1. Since $K = (K^*)^*$, every two-dimensional convex body is a polar of another convex body, and the result follows. □

Let us write the Levy representation in a little different form that looks more complicated than the original representation, but it will be useful in the future.

LEMMA 6.9. *Let K be an origin-symmetric star body in $\mathbb{R}^n$, $p > 0$, where p is not an even integer. The space $(\mathbb{R}^n, \|\cdot\|_K)$ embeds in L_p if and only if there exists a finite Borel measure μ on the sphere so that for every even test function $\phi \in \mathcal{S}(\mathbb{R}^n)$, whose Fourier transform is supported outside of the origin, we have*

$$\int_{\mathbb{R}^n} \|x\|_K^p \phi(x) dx = \frac{1}{\Gamma(-p/2)} \int_{S^{n-1}} \left(\int_{\mathbb{R}} |z|^{-1-p} \hat{\phi}(z\xi)\, dz \right) d\mu(\xi). \tag{6.4}$$

PROOF. By Lemma 6.4, the existence of an isometric embedding is equivalent to the existence of a measure μ satisfying the representation (6.2). Applying both sides of (6.2) to an even test function ϕ such that the Fourier transform $\hat{\phi}$ is supported outside of the origin, we get

$$\int_{\mathbb{R}^n} \|x\|^p \phi(x)\, dx = \int_{S^{n-1}} \left(\int_{\mathbb{R}^n} |(x,\xi)|^p \phi(x)\, dx \right) d\mu(\xi)$$

$$= \int_{S^{n-1}} \left(\int_{\mathbb{R}} |t|^p \left(\int_{(x,\xi)=t} \phi(x)\, dx \right) dt \right) d\mu(\xi) = \int_{S^{n-1}} \langle |t|^p, \mathcal{R}\phi(\xi; t)\rangle\, d\mu(\xi),$$

where $\mathcal{R}\phi(\xi; t)$ is the Radon transform of the function ϕ in the direction of ξ at the point t, and the action of $|t|^p$ on this Radon transform is in the sense of distributions. We now use the connection between the Radon and Fourier transforms, Lemma 2.11, and formula (2.40) for the Fourier transform of $|t|^p$ to conclude that

$$\int_{\mathbb{R}^n} \|x\|^p \phi(x)\, dx = \frac{1}{\Gamma(-p/2)} c_p \int_{S^{n-1}} \langle |z|^{-1-p}, \hat{\phi}(z\xi)\rangle\, d\mu(\xi),$$

where c_p is a positive constant, which can be included in the measure. Since $\hat{\phi}$ is supported outside of the origin, the action of the distribution $|z|^{-1-p}$ in the latter formula is by integration, and we get (6.4).

On the other hand, reversing the argument, we see that (6.4) implies that the distributions

$$\|\cdot\|^p \qquad \text{and} \qquad \int_{\mathbb{R}^n} |(\cdot, \xi)|^p d\mu(\xi)$$

coincide on test functions, whose Fourier transform is supported outside of the origin. Therefore, these functions are equal up to a polynomial. Since both of them are even and homogeneous of degree p which is not an even integer, the polynomial is equal to zero, and we get (6.2). $\square$

Comparing the formula of Lemma 6.9 with the result of the second part of Corollary 2.26, we get a Fourier analytic criterion of embeddability in L_p that applies to every $p > 0$ which is not an even integer.

THEOREM 6.10. *Let K be an origin-symmetric star body in $\mathbb{R}^n$, and assume the number $p > 0$ is not an even integer. Then the space $(\mathbb{R}^n, \|\cdot\|_K)$ embeds isometrically in L_p if and only if $\Gamma(-p/2)(\|\cdot\|_K^p)^\wedge$ is a positive distribution on $\mathbb{R}^n \setminus \{0\}$.*

6.2. Schoenberg's problems on positive definite functions

In this section we present another application of the connection between positive definite functions and isometric embeddings in L_p. The following problem was posed by Schoenberg in 1938 and was completely solved in the beginning of the 1990's: given numbers $n \geq 2$ and $q \in (2, \infty]$, for which $p \in (0, 2]$ is the function $\exp(-\|\cdot\|_q^p)$ positive definite on $\mathbb{R}^n$? Here $\|\cdot\|_q$ is the norm of the space ℓ_q^n. In view of Theorem 6.6, an equivalent formulation is: for which $p \in (0, 2)$ does the space ℓ_q^n, $q > 2$, embed isometrically in L_p?

The solution shows that the function $\exp(-\|\cdot\|_q^p)$ is not positive definite if $n \geq 3$, $0 < p \leq 2$ and $q > 2$. If $n = 2$ and $q > 2$, this function is positive definite if $0 < p \leq 1$ (by Theorem 6.6 and Corollary 6.8), and it is not positive definite if $1 < p \leq 2$. Note that for $p > 2$ the function $\exp(-|\cdot|^p)$ is not positive definite on $\mathbb{R}$ (see the remark in the very end of Section 2.8), so Schoenberg's question is of no interest for $p > 2$.

In the case $2 < q < \infty$ the original solution to Schoenberg's problem in [K1] is as follows. First, by Theorem 6.10, the space ℓ_q^n embeds in L_p, $0 < p \leq 2$, if and only if the distribution $(\|\cdot\|_q^p)^\wedge$ is negative on $\mathbb{R}^n \setminus \{0\}$. Now we use Lemma 3.6 to compute the Fourier transform of $\|\cdot\|_q^p$ (note that the formula of Lemma 3.6 works when $p < nq$ which is our case because $p \leq 2$ and $q > 2$). To prove that for $n = 3$, $0 < p \leq 2$ and $n = 2$, $1 < p \leq 2$, the space ℓ_q^n does not embed in L_p, we use exactly the same argument as in the proof of Lemma 4.12. In the case $n = 2$, $0 < p \leq 1$ the existence of an isometric embedding of ℓ_q^n in L_p follows from Corollaries 6.8 and 6.7.

We prove here a result generalizing the solution to Schoenberg's problem. This result is in the spirit of Theorem 4.19, but the proof is different, and also it is possible to remove one of the assumptions of Theorem 4.19. In the next theorem we consider three-dimensional spaces. However, a restriction of a positive definite function to a subspace is also positive definite, so the conclusion below holds for a space of any dimension containing a three-dimensional subspace satisfying the assumptions of the theorem.

THEOREM 6.11. *Let X be a three-dimensional normed space with a normalized basis $\{e_1, e_2, e_3\}$ so that the following hold.*

(i) For every fixed $(x_2, x_3) \in \mathbb{R}^2\setminus\{0\}$, the function $x_1 \mapsto \|x_1e_1 + x_2e_2 + x_3e_3\|$ has a continuous second derivative everywhere on $\mathbb{R}$, and

$$\|x\|'_{x_1}(0, x_2, x_3) = \|x\|''_{x_1^2}(0, x_2, x_3) = 0,$$

where $\|x\|'_{x_1}$ and $\|x\|''_{x_1^2}$ stand for the first and second derivatives by x_1, respectively.

(ii) There exists a constant C so that for every $x_1 \in \mathbb{R}$ and every $(x_2, x_3) \in \mathbb{R}^2$ with $\|x_2e_2 + x_3e_3\| = 1$, one has $\|x\|''_{x_1^2}(x_1, x_2, x_3) \le C$.

Then, for every $0 < p \le 2$, the function $\exp(-\|\cdot\|^p)$ is not positive definite, and the space X does not embed isometrically in L_p, $0 < p \le 2$.

PROOF. We start with a few remarks that are similar to the remarks preceding Lemma 4.20:

(i) As follows from Lemma 2.1, for any $p > 0$, the function $(x_2, x_3) \mapsto \|x_2e_2 + x_3e_3\|^{p-2}$ is locally integrable on $\mathbb{R}^2$.

(ii) A simple consequence of the triangle inequality is that $-1 \le \|x\|'_{x_1} \le 1$ at every point $x \in \mathbb{R}^3$ with $(x_2, x_3) \ne 0$.

(iii) For every fixed $(x_2, x_3) \in \mathbb{R}^2\setminus\{0\}$, $\|x\|$ is a convex differentiable function of x_1 whose derivative at zero is equal to zero. Therefore, for every $x = (x_1, x_2, x_3) \in \mathbb{R}^3$, we have $\|x\| \ge \|x_2e_2 + x_3e_3\|$.

(iv) The function $\|x\|''_{x_1^2}$ is non-negative, homogeneous of degree -1. Let C be the constant from condition (ii) of Theorem 6.11. Then, for every $x_1 \in \mathbb{R}$ and every $(x_2, x_3) \in \mathbb{R}^2 \setminus \{0\}$, the second derivative $\|x\|''_{x_1^2}(x_1, x_2, x_3)$ is equal to

$$\frac{1}{\|x_2e_2 + x_3e_3\|}\|x\|''_{x_1^2}\Big(\frac{x_1}{\|x_2e_2 + x_3e_3\|}, \frac{x_2}{\|x_2e_2 + x_3e_3\|}, \frac{x_3}{\|x_2e_2 + x_3e_3\|}\Big),$$

which is less than or equal to $C/\|x_2e_2 + x_3e_3\|$ by condition (ii) of Theorem 6.11.

By Corollary 6.7 and Theorem 6.6, it is enough to prove the theorem for $0 < p < 1$. Suppose that the function

$$F(x_1, x_2, x_3) = \exp(-\|x_1e_1 + x_2e_2 + x_3e_3\|^p)$$

is positive definite in $\mathbb{R}^3$. Similarly to [Za2], we pass to a positive definite function of one variable as follows. Let

$$h(x_2, x_3) = \begin{cases} (1 - |x_2|)(1 - |x_3|), & \text{if} \quad \max\{|x_2|, |x_3|\} \le 1, \\ 0, & \text{otherwise.} \end{cases}$$

Denote $d\mu(x_2, x_3) = h(x_2, x_3)\, dx_2\, dx_3$. Let us prove that the function

$$\psi(t) = \int_{\mathbb{R}^2} \exp(-\|te_1 + x_2e_2 + x_3e_3\|^p)\, d\mu(x_2, x_3) \tag{6.5}$$

is positive definite on $\mathbb{R}$. In fact, the function $h(x_2, x_3)$ is positive definite because

$$\int_{\mathbb{R}^2} e^{-i(u_2x_2+u_3x_3)} d\mu(x_2, x_3) = \left(\frac{2\sin(u_2/2)}{u_2}\right)^2 \left(\frac{2\sin(u_3/2)}{u_3}\right)^2 \ge 0.$$

Therefore, for each $\epsilon > 0$, the function $F(x_1, x_2, x_3)h(x_2, x_3)e^{-\epsilon|x_1|}$ is positive definite on $\mathbb{R}^3$, as the product of positive definite functions. Hence,

$$\int_{\mathbb{R}^3} e^{-i(u,x)} F(x_1, x_2, x_3)e^{-\epsilon|x_1|}\, dx_1\, d\mu(x_2, x_3) \ge 0, \qquad \forall (u_1, u_2, u_3) \in \mathbb{R}^3.$$

By putting $u_2 = u_3 = 0$, we get

$$\int_{-\infty}^{\infty} e^{-ix_1 u} \left(\int_{\mathbb{R}^2} F(x_1, x_2, x_3)\, d\mu(x_2, x_3) \right) e^{-\epsilon|x_1|}\, dx_1 \geq 0, \qquad u \in \mathbb{R}.$$

Therefore, the function $x_1 \mapsto e^{-\epsilon|x_1|}\psi(x_1)$ is positive definite. Since $\epsilon > 0$ is arbitrary, $\psi(t)$ is a positive definite function, as the limit of positive definite functions. This argument appeared in [Za2].

Now we differentiate twice the function under the integral in (6.5) and prove that $\psi''(0) = 0$.

Assume from now on that $(x_2, x_3) \in \mathbb{R}^2\backslash\{0\}$. We have

$$F'_{x_1}(x_1, x_2, x_3) = -p\|x_1e_1 + x_2e_2 + x_3e_3\|^{p-1}\|x\|'_{x_1} F(x_1, x_2, x_3).$$

By remarks (ii), (iii) we get (recall that $0 < p < 1$)

$$(6.6) \qquad \left|F'_{x_1}(x_1, x_2, x_3)\right| \leq p\|x_1e_1 + x_2e_2 + x_3e_3\|^{p-1} \leq p\|x_2e_2 + x_3e_3\|^{p-1},$$

for every $x_1 \in \mathbb{R}$ and $(x_2, x_3) \in \mathbb{R}^2\backslash\{0\}$. For the second derivative, we have

$$F''_{x_1^2}(x_1, x_2, x_3) = F(x_1, x_2, x_3)\big[-p(p-1)\|x\|^{p-2}(\|x\|'_{x_1})^2 - p\|x\|^{p-1}\|x\|''_{x_1^2}$$
$$+p^2\|x\|^{2(p-1)}(\|x\|'_{x_1})^2\big].$$

By remarks (ii), (iii), (iv), and since $0 < p < 1$ and $|F| \leq 1$, we have for every $x_1 \in \mathbb{R}$ and $(x_2, x_3) \in \mathbb{R}^2\backslash\{0\}$,

$$|F''_{x_1^2}(x_1, x_2, x_3)|$$
$$\leq (1-p)p\|x_1e_1 + x_2e_2 + x_3e_3\|^{p-2} + pC\|x_1e_1 + x_2e_2 + x_3e_3\|^{p-1}\|x_2e_2 + x_3e_3\|^{-1}$$
$$+p^2\|x_1e_1 + x_2e_2 + x_3e_3\|^{2p-2}$$
$$\leq p(C + 1 - p)\|x_2e_2 + x_3e_3\|^{p-2} + p^2\|x_2e_2 + x_3e_3\|^{2p-2}.$$

Thus both the first and second partial derivatives of F by the variable x_1 are bounded above by locally integrable functions in $\mathbb{R}^2$, which do not depend on x_1. (Here we use remark (i)). We have for every $t > 0, \epsilon \neq 0$,

$$\frac{\psi(t+\epsilon) - \psi(t)}{\epsilon} = \int_{\mathbb{R}^2\backslash\{0\}} \frac{F(t+\epsilon, x_2, x_3) - F(t, x_2, x_3)}{\epsilon} d\mu(x_2, x_3),$$

$$= \int_{\mathbb{R}^2\backslash\{0\}} F'_t(\theta(t, \epsilon, x_2, x_3), x_2, x_3) d\mu(x_2, x_3),$$

where $\theta(t, \epsilon, x_2, x_3) \in (t, t+\epsilon)$ (or $(t-\epsilon, t)$). By (6.6), the function $|F'_{x_1}(x_1, x_2, x_3)|$ is bounded by an integrable function of the variables x_2, x_3, so we can apply Lebesgue's bounded convergence theorem and pass to the limit as $\epsilon \to 0$, under the integral sign. By continuity of the first derivative we get

$$\psi'(t) = \int_{\mathbb{R}^2\backslash\{0\}} F'_t(t, x_2, x_3) d\mu(x_2, x_3).$$

Repeating this argument, we obtain

$$\psi''(t) = \int_{\mathbb{R}^2\backslash\{0\}} F''_{t^2}(t, x_2, x_3) d\mu(x_2, x_3).$$

By condition (i) of Theorem 6.11, the integrands in the last two integrals tend pointwise to zero as $t \to 0$. These integrands are also bounded above by integrable functions independent of t. Therefore, applying Lebesgue's bounded convergence theorem once more, we finally get $\psi'(0) = \psi''(0) = 0$. Hence, $\psi(t) = \psi(0) + o(t^2)$ as

$t \to 0$. Since ψ is positive definite, we must have $\psi(t) \equiv \psi(0)$ (see [Luc, Th. 4.1.1]; we mentioned this property of positive definite functions just before Theorem 4.19). In other words

$$\int_{\mathbb{R}^2} F(t, x_2, x_3) d\mu(x_2, x_3) \equiv \int_{\mathbb{R}^2} F(0, x_2, x_3) d\mu(x_2, x_3), \qquad \forall t.$$

Since $F(t, x_2, x_3) \leq F(0, x_2, x_3)$ (remark (iii)) and since μ is a positive measure with continuous density, we must have $F(t, x_2, x_3) \equiv F(0, x_2, x_3)$ for every t and (x_2, x_3) such that $\max\{|x_2|, |x_3|\} \leq 1$, and this is clearly a contradiction. $\square$

Note that for $n = 2$ the statement of Theorem 6.11 is not true, because every two-dimensional space embeds isometrically in L_p for every $p \in (0, 1]$. The reason for why the proof of Theorem 6.11 does not work for two-dimensional spaces is that, unlike the function $\|x_2 e_2 + x_3 e_3\|^{p-2}$ on $\mathbb{R}^2$, the function $|x_2|^{p-2}$ is not locally integrable on $\mathbb{R}$ if $0 < p \leq 1$.

However, the proof above can be used to show that two-dimensional spaces satisfying condition (i) of Theorem 6.11 do not embed in L_p with $1 < p < 2$. There is, however, a simpler argument. Indeed, if a space embeds in L_p with $1 < p < 2$, then, by Lemma 6.4, there exists a finite Borel measure μ on the sphere S^1 so that, for every $x \in \mathbb{R}^2$,

$$\|x\|^p = \int_{S^1} |(x, \xi)|^p d\mu(\xi).$$

Differentiating both sides twice by x_1 (which is possible now because $p > 1$), we get that

$$p(p-1)\|x\|^{p-2}(\|x\|'_{x_1})^2 + p\|x\|^{p-1}\|x\|''_{x_1^2} = \int_{S^1} |(x, \xi)|^{p-2} \xi_1^2 d\mu(\xi).$$

Putting $x_1 = 0$ and using condition (i) of Theorem 6.11, we get a contradiction. The fact that the spaces ℓ_q^2, $q > 2$, do not embed in L_p with $1 < p < 2$ was essentially proved in [Do] and also in [K1] and [GrM].

It is easy to check that the spaces ℓ_q^n, $2 < q < \infty$, satisfy the conditions of Theorem 6.11, so the solution to Schoenberg's problem is a particular case of this theorem. One can also formulate more general results for q-sums and for Orlicz spaces, as in Theorems 4.21 and 4.22.

COROLLARY 6.12. *If $n \geq 3$ and $2 < q < \infty$, then the space ℓ_q^n does not embed isometrically in L_p for any $p \in (0, 2]$.*

In the case $q = \infty$ Schoenberg's problem was solved by Misiewicz [Mis2]. As it was done in [Mis2], we prove a more general result. The proof presented here comes from [K4].

THEOREM 6.13. *If $n \geq 3$, f is an even continuous function on $\mathbb{R}$, $f(0) = 1$ and the function $f(\|\cdot\|_\infty)$ is positive definite on $\mathbb{R}^n$, then $f \equiv 1$.*

PROOF. It suffices to prove the theorem for $n = 3$ because the restriction of a positive definite function to a subspace is also positive definite.

By Bochner's theorem, Theorem 6.2, there exists a probability measure μ on $\mathbb{R}^3$ such that $\widehat{\mu} = f(\|\cdot\|_\infty)$, and there also exists a probability measure ν on $\mathbb{R}$ such that $\widehat{\nu} = f$.

We are going to use Lemma 4.16 with the measure ν in place of the function u. The proof is exactly the same, and we also no longer need to restrict ourselves to test functions supported outside of the hyperplanes $(\delta, x) = 0$. We needed this in Lemma 4.16 because we aimed at functions u discontinuous at zero and not necessarily locally integrable around zero.

Let $\phi \in S(\mathbb{R}^3)$ be a non-negative function with compact support in $\mathbb{R}_3^+ = \{\xi \in \mathbb{R}^3 : \ \xi_k > 0, \ 1 \le k \le 3\}$. By Lemma 4.16 applied to $n = 3$ and ν in place of u,

$$0 \le (2\pi)^3 \int_{\mathbb{R}^3} \phi(x) \ d\mu(x) = \langle f(\|\cdot\|_\infty)^\wedge, \phi\rangle$$

$$= -\pi \int_R \left(\sum_\delta \delta_1\delta_2\delta_3 \int_{(\delta,x)=y} \frac{\phi(x)dx}{x_1x_2x_3} \right) yd\nu(y). \tag{6.7}$$

We used the fact that, in the sense of distributions,

$$\widehat{f} = (\hat{\nu})^\wedge = 2\pi\nu \quad \text{and} \quad f(\|\cdot\|_\infty)^\wedge = (\hat{\mu})^\wedge = (2\pi)^3\mu,$$

because ν and μ are even distributions on $\mathbb{R}$ and $\mathbb{R}^n$, respectively.

The set of functions ϕ with properties described above is dense in the space of continuous functions having compact support in the closure of $\mathbb{R}_3^+$ and vanishing in the coordinate planes. Hence, (6.7) remains valid for every function of the form $\phi(x) = x_1x_2x_3\phi_1(x_1)\phi_2(x_2)\phi_3(x_3)$ with ϕ_1, ϕ_2, ϕ_3 being continuous functions on $[0, \infty)$ with compact support.

Fix numbers t_1, t_2 with $0 \le t_2 \le t_1 - 1$. Let $\{\omega_n\}_{n=1}^\infty$ be a sequence of non-negative continuous functions on $\mathbb{R}$ such that $\int_\mathbb{R} \omega_n(x)dx = 1$ and $\text{supp}(\omega_n) \subset (0, 1/n)$ for every $n \in \mathbb{N}$. Let $\psi_{1,n}(t) = \omega_n(t - t_1), \psi_{2,n}(t) = \omega_n(t - t_2)$ and let ψ_3 be the characteristic function of the segment $[0, 1]$.

Put $\phi_n(x) = x_1x_2x_3\psi_{1,n}(x_1)\psi_{2,n}(x_2)\psi_3(x_3)$. As $n \to \infty$, applying (6.7) to the functions ϕ_n, we get

$$\int_{t_1+t_2}^{t_1+t_2+1} yd\nu(y) - \int_{t_1+t_2-1}^{t_1+t_2} yd\nu(y)$$
$$- \int_{t_1-t_2}^{t_1-t_2+1} yd\nu(y) + \int_{t_1-t_2-1}^{t_1-t_2} yd\nu(y) \le 0.$$

Let $n \in \mathbb{N} \cup \{0\}, k \in \mathbb{N}, t_1 - t_2 = n + 1, t_1 + t_2 = n + k$. Then we have

$$\int_{n+k}^{n+k+1} yd\nu(y) - \int_{n+k-1}^{n+k} yd\nu(y) \le \int_{n+1}^{n+2} yd\nu(y) - \int_n^{n+1} yd\nu(y). \tag{6.8}$$

If there exists a number $n \in \mathbb{N} \cup \{0\}$ for which the right-hand side of (6.8) is negative, say less than $\delta < 0$, then adding the inequalities (6.8) with $k = 1, ..., m$ we get that, for any $m \in \mathbb{N}$,

$$\int_{n+m}^{n+m+1} yd\nu(y) - \int_n^{n+1} yd\nu(y) \le m\delta.$$

For large m this implies $\int_{n+m}^{n+m+1} yd\nu(y) < 0$ which is impossible. Thus, the left-hand side of (6.8) is non-negative for every n, so the sequence $\int_n^{n+1} yd\nu(y)$, $n \in \mathbb{N} \cup \{0\}$, is non-decreasing.

Now if there exists n_0 so that $\int_{n_0}^{n_0+1} yd\nu(y) > c > 0$, then, for every $n \ge n_0$, we have

$$(n + 1)\nu([n, n + 1]) \ge \int_n^{n+1} yd\nu(y) > c.$$

Therefore, $\nu([0,+\infty)) = \infty$ and ν is not a probability measure. We get a contradiction.

We are left with the only option that $\int_n^{n+1} y d\nu(y) = 0$ for all $n \in \mathbb{N} \cup \{0\}$. In this case ν is the unit mass at zero and $f \equiv 1$, which proves the theorem. □

6.3. Intersection bodies and embeddings in L_p, $p < 0$

In this section we extend the concept of isometric embedding into L_p to negative values of p. Embedding in L_{-p} is just a concept; we do not pretend to introduce L_{-p} as any kind of function space. The advantage of this concept is in its connection with k-intersection bodies. It appears that a space $(\mathbb{R}^n, \|\cdot\|_K)$ embeds in L_{-k} if and only if the body K is a k-intersection body. This connection opens a way of getting new geometric results by extending to negative values of p different facts from the well-developed theory of L_p-spaces with $p > 0$. A brief description of these results along with motivation for introducing embeddings in L_{-p} was given in the Introduction.

In order to introduce embedding in L_p with $p < 0$, let us look again at the result of Lemma 6.9: for an origin-symmetric star body D, the space $(\mathbb{R}^n, \|\cdot\|_D)$ embeds in L_p, $p > 0$, $p \notin 2\mathbb{N}$, if and only if there exists a finite Borel measure μ on the sphere so that for every even test function ϕ,

$$\int_{\mathbb{R}^n} \|x\|_D^p \phi(x)dx = \frac{1}{\Gamma(-p/2)} \int_{S^{n-1}} \langle |z|^{-1-p}, \hat{\phi}(z\xi)\rangle d\mu(\xi). \tag{6.9}$$

On the other hand, let us combine the results of Theorem 4.8 and Corollary 2.26: an origin-symmetric star body D is a k-intersection body if and only if $\|\cdot\|_D^{-k}$ is a positive definite distribution, which, in turn, is equivalent to the existence of a finite Borel measure μ on the sphere such that for every even test function ϕ,

$$\int_{\mathbb{R}^n} \|x\|_D^{-k} \phi(x)dx = \int_{S^{n-1}} \left(\int_0^\infty z^{k-1}\hat{\phi}(z\xi)\, dz \right) d\mu(\xi)$$
$$= \frac{1}{2} \int_{S^{n-1}} \langle |z|^{k-1}, \hat{\phi}(z\xi)\rangle d\mu(\xi).$$

We see that the latter equality is the same as (6.9) with $p = -k$. These considerations lead to the definition of embedding in L_{-p}.

DEFINITION 6.14. *Let D be an origin-symmetric star body in $\mathbb{R}^n$. We say that the space $(\mathbb{R}^n, \|\cdot\|_D)$ embeds in L_{-p}, where $0 < p < n$, if there exists a finite Borel measure μ on S^{n-1} so that for every even test function ϕ*

$$\int_{\mathbb{R}^n} \|x\|_D^{-p} \phi(x)\, dx = \int_{S^{n-1}} \left(\int_0^\infty t^{p-1}\hat{\phi}(t\xi)\, dt \right) d\mu(\xi). \tag{6.10}$$

An immediate consequence of this definition and Corollary 2.26 is the following extension to negative values of p of the Fourier characterization of embeddings in L_p, $p > 0$, Theorem 6.10.

THEOREM 6.15. *Let D be an origin-symmetric star body in $\mathbb{R}^n$ and let $p \in (0,n)$. The space $(\mathbb{R}^n, \|\cdot\|_D)$ embeds in L_{-p} if and only if $\|\cdot\|_D^{-p}$ represents a positive definite distribution.*

Summarizing the remarks above, we see that k-intersection bodies can be considered as the unit balls of subspaces of L_{-k} :

THEOREM 6.16. *The following are equivalent:*
(i) An origin-symmetric star body D is a k-intersection body.
(ii) $\|\cdot\|_D^{-k}$ is a positive definite distribution.
(iii) The space $(\mathbb{R}^n, \|\cdot\|_D)$ embeds in L_{-k}.

Let us show several applications.

Example 1. A well-known simple fact (see Corollary 6.8) is that every two-dimensional normed space embeds in L_1. How can this fact be extended to negative p?

By Corollary 4.9, for every n-dimensional normed space $(\mathbb{R}^n, \|\cdot\|)$, $n \geq 4$, the function $\|\cdot\|^{-n+3}$ represents a positive definite distribution. In view of Theorem 6.16, this means that every n-dimensional normed space embeds in L_{-n+3}.

If $n = 2$, we get the two-dimensional fact mentioned above. Putting $n = 4$, we get that every four-dimensional normed space embeds in L_{-1} or, in other words, every origin-symmetric convex body in $\mathbb{R}^4$ is an intersection body. As we already know, this fact is crucial for the affirmative part of the solution to the Busemann-Petty problem.

The result also holds for $n = 3$, where every three-dimensional normed space embeds in L_0. The concept of embedding in L_0 was introduced in [KKYY] as follows: a space $(\mathbb{R}^n, \|\cdot\|)$ embeds in L_0 if there exist a finite Borel measure μ on S^{n-1} and a constant C so that for every $x \in \mathbb{R}^n$

$$\ln \|x\| = \int_{S^{n-1}} \ln |(x, \xi)| \, d\mu(\xi) + C.$$

It was proved in [KKYY] that a normed space embeds in L_0 if and only if the Fourier transform of $\ln(\|\cdot\|)$ is a negative distribution outside of the origin. This is true for every three-dimensional normed space, as shown in [KKYY] by extending the formula of Theorem 3.18 to the case $q = n - 1$.

Example 2. Another well-known fact from the isometric theory of L_p-spaces (see Corollary 6.7) is that L_q embeds isometrically in L_p for every $0 < p < q \leq 2$. This means that L_p-spaces become larger when p decreases from 2 to 0. One can expect L_p to be even larger when $p < 0$. Our next result proves that this is indeed the case. The proof is the same as that of Theorem 4.11, so we omit some details.

THEOREM 6.17. *Every n-dimensional subspace of L_q, $0 < q \leq 2$, embeds in L_{-p} for every $0 < p < n$. In particular, the unit ball of any n-dimensional subspace of L_q with $0 < q \leq 2$ is a k-intersection body for every $0 < k < n$.*

PROOF. Let D be the unit ball of an n-dimensional subspace of L_q. By Theorem 6.6, $\exp(-\|\cdot\|^q)$ is a positive definite function. Then for every $t > 0$ the function $\exp(-t^q\|\cdot\|^q)$ is also positive definite. For every $p \in (0, n)$, writing

$$\|x\|_D^{-p} = \frac{q}{\Gamma(p/q)} \int_0^\infty t^{p-1} \exp(-t^q \|x\|_D^q) \, dt$$

we see that $\|\cdot\|_D^{-p}$ is a positive combination of positive definite functions. Now, applying both sides of the latter equation to an even test function ϕ whose Fourier transform is non-negative everywhere on $\mathbb{R}^n$, we see that positive definiteness of the distribution $\|\cdot\|_D^{-p}$ follows from the Fubini theorem. □

The result of Theorem 6.17, in conjunction with Theorem 5.2, confirms the fact that the answer to the question of the Busemann-Petty problem is affirmative if K

is the unit ball of a subspace of L_q, where $0 < q \leq 2$ and L is any origin-symmetric star body.

Example 3. Recall that the solution to Schoenberg's problem, Corollary 6.12, shows that, for $n \geq 3$, the space l_q^n, $q > 2$, does not embed in any L_p, $0 < p < 2$. For $n = 2$, the embedding exists if and only if $0 < p \leq 1$.

It follows from Theorems 6.15 and 4.13 that, for $2 < q < \infty$, the space ℓ_q^n does not embed in L_{-p} if $p < n - 3$. Clearly, this fact extends to negative p the solution to Schoenberg's problem. In particular, the space ℓ_q^5 does not embed in L_{-1}, which means that the unit ball of the space ℓ_q^5 is not an intersection body. This result provides a counterexample to the Busemann-Petty problem in dimension 5.

In the rest of this section we show an example of an isomorphic property of L_p-spaces that can be extended to negative values of p. The result from the theory of L_p-spaces that we extend is the factorization theorem of Maurey and Nikishin [Ma], [Ni], [W], which, in particular, states that if $0 < p < q < 1$, then every Banach subspace of L_p is isomorphic to a subspace of L_q. We prove that, for any $-\infty < p < q < 1$, $p \neq 0$, $q > 0$, every n-dimensional Banach subspace of L_p with $-n < p$ is isomorphic to a subspace of L_q with the Banach-Mazur distance depending only on p and q (see Theorem 6.26 below). Recall that the *Banach-Mazur distance* between Banach spaces X and Y is defined by

$$d(X, Y) = \inf\{\|T\|\|T^{-1}\| : \ T : X \to Y \quad \text{is a linear isomorphism}\}.$$

In other words, $d(X, Y)$ is the infimum of all numbers C such that there exists a linear bijection $T : X \to Y$ satisfying

$$\|x\|_X \leq \|Tx\|_Y \leq C\|x\|_X$$

for every $x \in X$.

In terms of intersection bodies this result translates as follows:

THEOREM 6.18. *For any $k \in \mathbb{N}$ and $0 < q < 1$, there exists a constant $c(k, q)$ depending on k and q only so that for every $n \in \mathbb{N}$ and every origin-symmetric convex k-intersection body D in $\mathbb{R}^n$, $n > k$, there exists an n-dimensional subspace of $L_q([0, 1])$ whose unit ball L satisfies $L \subset D \subset c(k, q)L$.*

The latter theorem, in conjunction with the fact that the unit ball of any subspace of L_q, $0 < q < 1$, is a k-intersection body for every k, Theorem 6.17, shows that origin-symmetric convex k-intersection bodies are isomorphically equivalent to unit balls of subspaces of L_q, $0 < q < 1$.

If $k = 1$ and D is the 1-intersection body of an origin-symmetric convex body, one can choose L in Theorem 6.18 as an ellipsoid. This follows from the result of Hensley, Corollary 2.7, that there exist absolute constants a_1, a_2 so that every origin-symmetric convex body D in $\mathbb{R}^n$ admits a linear transformation TD satisfying

$$a_1 \leq \frac{\mathrm{Vol}_{n-1}(TD \cap \xi^\perp)}{\mathrm{Vol}_{n-1}(TD \cap \eta^\perp)} \leq a_2$$

for any $\xi, \eta \in S^{n-1}$. However, as shown in Section 4.3, the most interesting and useful examples of 1-intersection bodies are not the ones generated by convex bodies, but rather by star bodies or measures on the sphere (in the latter case they belong to the closure in the radial metric of the class of intersection bodies of star bodies). For example, the cube in $\mathbb{R}^4$ is a 1-intersection body generated by a measure on the sphere whose density is not bounded, as follows from the formula for the Fourier transform of $\|\cdot\|_\infty^{-1}$, Lemma 4.17.

We now start preparations for the proof of Theorem 6.26. First, we need a simple comparison principle:

LEMMA 6.19. *If f, g are two positive random variables so that, for any $t \geq 0$, the expectation $\mathbb{E}(e^{-tf}) \leq \mathbb{E}(e^{-tg})$, then $(\mathbb{E}f^p)^{\frac{1}{p}} \geq (\mathbb{E}g^p)^{\frac{1}{p}}$ whenever $-\infty < p \leq 1$.*

PROOF. From the definition of the Γ-function (see Section 2.4)

$$x^z\Gamma(-z) = \int_0^\infty t^{-z-1}e^{-xt}\,dt, \qquad \Re z < 0. \tag{6.11}$$

Then if f is a non-negative random variable for $\Re z < 0$, we have

$$\mathbb{E}(f^z) = \frac{1}{\Gamma(-z)}\int_0^\infty t^{-z-1}\mathbb{E}(e^{-tf})\,dt. \tag{6.12}$$

Now let us note that by analytic continuation (as in (2.37)), equality (6.11) implies

$$-\frac{1}{z} - x^z\Gamma(z) = \int_0^\infty t^{-z-1}(\chi_{[0,1]}(t) - e^{-xt})\,dt,$$

whenever $\Re z < 1$, and this implies that if $0 < \Re z < 1$,

$$-x^z\Gamma(-z) = \int_0^\infty t^{-z-1}(1 - e^{-xt})\,dt. \tag{6.13}$$

The statement of the lemma easily follows from these equalities. □

Let $(\eta_k)_{k=1}^\infty$ be a sequence of independent normalized symmetric q-stable random variables, $0 < q \leq 2$; see Section 2.8 for the definition. By Lemma 6.3, any linear combination $\sum a_k\eta_k$ has the same distribution as $\|a\|_q\eta_1$. Hence, by formula (2.55), for $-1 < p < q$, $p \neq 0$, the expectation

$$\mathbb{E}(|\sum_{k=1}^m a_k\eta_k|^p) = \|a\|_q^p\mathbb{E}(|\eta_1|^p) = \|a\|_q^p\frac{2^{p+1}\Gamma(-p/q)\Gamma((p+1)/2)}{q\sqrt{\pi}\Gamma(-p/2)}.$$

It appears that, for $0 < q < 1$, if we replace η_k's by their absolute values, the expectation increases by no more than a scalar multiple not depending on a, and also we get an upper bound for all $-\infty < p < q$, $p \neq 0$.

We use the fact that the density of the probability distribution of a normalized symmetric 1-stable random variable can be computed precisely and is equal to the function $1/(\pi(1+x^2))$ (the Cauchy distribution), which can be checked by a direct computation of the Fourier transform of the function $\exp(-|x|)$.

LEMMA 6.20. *Suppose $-\infty < p < q < 1$ with $p \neq 0$ and $q > 0$. Suppose $(\eta_k)_{k=1}^\infty$ is a sequence of independent normalized symmetric q-stable random variables. Then for any $a_1, \ldots, a_n \in \mathbb{R}$,*

$$(\mathbb{E}(\sum_{k=1}^n |a_k||\eta_k|)^p)^{\frac{1}{p}} \leq \left(\cos(\frac{q\pi}{2})\right)^{-\frac{1}{q}}\left(\frac{\Gamma(\frac{-p}{q})}{q\Gamma(-p)}\right)^{\frac{1}{p}}(\sum_{k=1}^n |a_k|^q)^{\frac{1}{q}}.$$

PROOF. Let $f = \sum_{k=1}^{n} |a_k||\eta_k|$. Then, for $t \geq 0$, by independence,

$$\mathbb{E}(e^{-tf}) = \prod_{k=1}^{n} \mathbb{E}(e^{-t|a_k||\eta_k|}).$$

Now

$$\mathbb{E}(e^{-t|a_k||\eta_k|}) = \mathbb{E}_\eta \mathbb{E}_\zeta (e^{-ita_k\eta_k\zeta})$$

where ζ is a normalized symmetric 1-stable random variable independent of η_k. Hence, computing first the expectation with respect to η_k,

$$\mathbb{E}(e^{-t|a_k||\eta_k|}) = \mathbb{E}_\zeta(e^{-t^q|a_k|^q|\zeta|^q}) = \frac{1}{\pi}\int_{-\infty}^{\infty} \frac{e^{-t^q|a_k|^q|x|^q}}{1+x^2}dx. \tag{6.14}$$

Now, using the convexity of the function e^{-x}, the formula for the 1-stable density,

$$\frac{1}{\pi}\int_{-\infty}^{\infty} \frac{dx}{1+x^2} = 1,$$

and Lemma 2.16 giving the expression for the Euler integral, we get that the quantity in (6.14) is greater than or equal to

$$\exp\left\{-\frac{|t|^q|a_k|^q}{\pi}\int_{-\infty}^{\infty}\frac{|x|^q}{1+x^2}dx\right\} = e^{-(\cos(\frac{q\pi}{2}))^{-1}t^q|a_k|^q}.$$

We thus have (recall that $q < 1$)

$$\mathbb{E}(e^{-tf}) \geq e^{-(\cos(\frac{q\pi}{2}))^{-1}t^q\sum_{k=1}^n|a_k|^q} = \mathbb{E}e^{-t(\cos(\frac{q\pi}{2}))^{-\frac{1}{q}}(\sum_{k=1}^n|a_k|^q)^{\frac{1}{q}}\alpha},$$

where α is a normalized positive q-stable random variable. By Lemma 6.19,

$$(\mathbb{E}(f^p))^{\frac{1}{p}} \leq (\cos(\frac{q\pi}{2}))^{-\frac{1}{q}}(\sum_{k=1}^{n}|a_k|^q)^{\frac{1}{q}}(\mathbb{E}(\alpha^p))^{\frac{1}{p}}.$$

This implies the conclusion by (2.53). □

Let Ω be a complete separable metric space. We denote by $\mathcal{M}(\Omega)$ the space of all real-valued Borel functions on Ω. Let μ be a Borel probability measure on Ω, i.e. $\mu(\Omega) = 1$. For any $f \in \mathcal{M}(\Omega)$ and any $-\infty < p < \infty$ with $p \neq 0$ we define

$$\|f\|_{p,d\mu} = \left(\int |f|^p d\mu\right)^{\frac{1}{p}}.$$

Notice that $0 \leq \|f\|_{p,d\mu} < \infty$ for all f and if $p < 0$, then $\|f\|_{p,d\mu} = 0$ does not imply $f = 0$ a.e.

We need an elementary extension to all p of a well-known property of the standard L_p-norms. We refer the reader to the book [KPR] for more information about L_p-spaces with $0 < p < 1$.

LEMMA 6.21. *Suppose $-\infty < p < q < \infty$, $p \neq 0$, $q \neq 0$. Then for any measurable functions $f_1, \ldots, f_n$ we have*

$$\|(\sum_{k=1}^{n}|f_k|^q)^{\frac{1}{q}}\|_{p,d\mu} \geq (\sum_{k=1}^{n}\|f_k\|_{p,d\mu}^q)^{\frac{1}{q}}.$$

PROOF. If $q > 0$, then $q/p > 1$ or $q/p < 0$, and the function $t^{q/p}$ is convex on $(0, \infty)$. Also, μ is a probability measure, so for any non-negative measurable function u,

$$\left(\int u \, d\mu\right)^{q/p} \leq \int u^{q/p} \, d\mu. \tag{6.15}$$

Putting $u = |f|^p$, we get that, for any measurable function f,

$$\|f\|_{p,d\mu} \leq \|f\|_{q,d\mu}. \tag{6.16}$$

If $q < 0$, then the function $t^{q/p}$ is concave, so inequality (6.15) reverses, but inequality (6.16) stays, because we raise both sides of (6.15) to a negative power $1/q$.

Let $F = (\sum_{k=1}^n |f_k|^q)^{1/q}$. Because of homogeneity in the desired inequality, we can assume that $\|F\|_{p,d\mu} = 1$. Then the measure ν defined by $d\nu = F^p d\mu$ is a probability measure. Put $g_k = |f_k|/F$. Clearly, $\sum_{k=1}^n g_k^q \equiv 1$, so, by (6.16) applied to the probability measure ν,

$$1 = \|(\sum_{k=1}^n g_k^q)^{1/q}\|_{q,d\nu} = (\sum_{k=1}^n \|g_k\|_{q,d\nu}^q)^{1/q} \geq (\sum_{k=1}^n \|g_k\|_{p,d\nu}^q)^{1/q}.$$

To finish the proof, just note that $\|g_k\|_{p,d\nu} = \|g_k F\|_{p,d\mu} = \|f_k\|_{p,d\mu}$. □

Let $X = (\mathbb{R}^n, \|\cdot\|_X)$ be an n-dimensional normed space, $\{e_k\}_{k=1}^n$ the standard basis in $\mathbb{R}^n$, and μ a Borel measure on S^{n-1}. We construct an embedding $U : X \mapsto \mathcal{M}(S^{n-1}, \mu)$ as follows: for every $x = \sum_{k=1}^n x_i e_i$ put $Ux(u) = (x, u)$, $u \in S^{n-1}$.

The next lemma generalizes to negative p a well-known connection between stable random variables and L_p-norms. The equality of this lemma can serve as a definition of embedding in L_p with negative p.

LEMMA 6.22. *Suppose that $-n < p < 0$ and that $X = (\mathbb{R}^n, \|\cdot\|_X)$ is an n-dimensional normed space embedding in L_p with the corresponding measure μ. If $\eta_1, ..., \eta_m$ are independent symmetric q-stable random variables, $0 < q \leq 2$, then, for any $f_1, ..., f_m \in X$ with the dimension of their linear span not smaller than n,*

$$\mathbb{E}\|\sum_{k=1}^m \eta_k f_k\|_X^p = c(p,q) \int_{S^{n-1}} (\sum_{k=1}^m |Uf_k(u)|^q)^{\frac{p}{q}} d\mu(u),$$

where $c(p,q)$ is a positive constant depending only on p and q.

PROOF. First consider the case $q = 2$, where η_k are symmetric Gaussians. Then the distribution of $\sum_{k=1}^m \eta_k f_k$ is a Gaussian measure in $\mathbb{R}^n$ with density $\phi \in \mathcal{S}(\mathbb{R}^n)$, whose characteristic function

$$\hat{\phi}(u) = \int_{\mathbb{R}^n} \phi(x) e^{-i(x,u)} \, dx$$

$$= \mathbb{E}(e^{-i(\sum_{k=1}^m \eta_k f_k, u)}) = \prod_{k=1}^m \mathbb{E}(e^{-i\eta_k (f_k, u)})$$

$$= \exp(-\frac{1}{2} \sum_{k=1}^m (f_k, u)^2).$$

Since ϕ is a test function, by the definition of embedding in L_p (formula (6.10)),

$$\mathbb{E}\|\sum_{k=1}^{m}\eta_k f_k\|_X^p = \int_{\mathbb{R}^n}\|x\|_X^p\phi(x)\,dx$$

$$= \int_{S^{n-1}}\left(\int_{\mathbb{R}}|t|^{-1-p}\hat{\phi}(tu)\,dt\right)d\mu(u)$$

$$= \int_{S^{n-1}}\left(\int_{\mathbb{R}}|t|^{-1-p}\exp(-\frac{t^2}{2}\sum_{k=1}^{m}(f_k,u)^2)\,dt\right)d\mu(u)$$

$$= c(p)\int_{S^{n-1}}\left(\sum_{k=1}^{m}(f_k,u)^2\right)^{\frac{p}{2}}d\mu(u),$$

which proves the case $q=2$ of the lemma.

Now, for arbitrary q, use the fact that each symmetric q-stable random variable η_k has the same distribution as $\sqrt{2\alpha_k}\gamma_k$ where α_k, γ_k are independent, γ_k is a normalized Gaussian and α_k is a normalized positive $q/2$-stable random variable; see the proof of Lemma 2.29. Using the case $q=2$ of Lemma 6.22, we get that

$$\mathbb{E}\|\sum_{k=1}^{m}\eta_k f_k\|_X^p = \mathbb{E}_\alpha\mathbb{E}_\gamma\|\sum_{k=1}^{n}\sqrt{2\alpha_k}\gamma_k f_k\|_X^p$$

$$= c(p)\mathbb{E}\left(\int_{S^{n-1}}\left(\sum_{k=1}^{m}\alpha_k(f_k,u)^2\right)^{\frac{p}{2}}d\mu(u)\right). \tag{6.17}$$

Note that $\sum_{k=1}^m\alpha_k(f_k,u)^2$ has the same distribution as $\alpha_1\left(\sum_{k=1}^m|(f_k,u)|^q\right)^{\frac{2}{q}}$. In fact, since α_k are independent positive $q/2$-stable, for every $t\in\mathbb{R}$, we have

$$\mathbb{E}(e^{-t\sum_{k=1}^m\alpha_k(f_k,u)^2}) = \prod_{k=1}^{m}\mathbb{E}(e^{-t\alpha_k(f_k,u)^2})$$

$$= e^{-|t|^{q/2}\sum_{k=1}^m|(f_k,u)|^q} = \mathbb{E}(e^{-t(\sum_{k=1}^m|(f_k,u)|^q)^{2/q}\alpha_1}),$$

and we can use the uniqueness theorem for the Laplace transform.

Now the expectation in (6.17) becomes

$$\mathbb{E}\alpha_1^{\frac{p}{q}}\int_{S^{n-1}}\left(\sum_{k=1}^{m}|(f_k,u)|^q\right)^{\frac{p}{q}}d\mu(u),$$

which proves the lemma. The constant $c(p,q)$ can be computed precisely using (2.26) and (2.53). □

The argument that we used in the case $q=2$ of Lemma 6.22 can be applied to all $0<q\le 2$. The problem is that, for $q<2$, the corresponding density ϕ is not a test function, and one cannot substitute it in the definition of embedding in L_p directly. It is, however, quite easy to make this argument work approximating ϕ by test functions. Then exactly the same argument works for all $q\in(0,2]$.

Our next lemma is a slight extension of the Komlos theorem [Kom]. For the proof, we refer the reader to [Kom] or [W, Theorem III.A.29] (where the result is more general) and then to [KK2] for a way to adjust the result to measurable functions that are allowed to assume the value $+\infty$ (we call these functions extended-valued).

LEMMA 6.23. *Suppose $(f_n)_{n=1}^{\infty}$ is a sequence of extended-valued measurable functions. Then there is a sequence (g_n) such that each g_n belongs to the convex hull $\mathrm{co}\{f_k : k \geq n\}$ and such that (g_n) converges a.e. to an extended-valued measurable function g.*

Our next theorem extends to negative p the classical change of density result of Maurey (see [Ma], [Pi1] or [W, p. 264]). This result gives an opportunity to reverse (in a certain way) inequality (6.16). In general, such an inversion is not possible even with adding a constant multiple independent of f. However, given a set V of measurable functions, if the L_p-norms are bounded for a much larger class of functions (q-convex combinations of functions from V), then L_q-norms of functions from V are also bounded after a certain change of density in the measure space.

THEOREM 6.24. *Suppose $-\infty < p < q < \infty$ with $q > 0$, $p \neq 0$, and suppose $V \subset \mathcal{M}(\Omega)$. Let*

$$\tilde{V}_q = \{(\sum_{j=1}^{m} c_j |f_j|^q)^{\frac{1}{q}} : \sum_{j=1}^{m} c_j = 1, 0 \leq c_j \leq 1; f_1, \dots, f_m \in V, m \in \mathbb{N}\}.$$

Suppose that there exists $h \in \tilde{V}_q$ with $\|h\|_{p,d\mu} > 0$. Then in order that there exist $v \in L_1(d\mu)$ with $v \geq 0$ a.e. and $\int v d\mu = 1$ such that

$$\|fv^{-\frac{1}{p}}\|_{q,vd\mu} \leq \sigma, \qquad \forall f \in V, \tag{6.18}$$

it is necessary and sufficient that

$$\|g\|_{p,d\mu} \leq \sigma, \qquad \forall g \in \tilde{V}_q. \tag{6.19}$$

PROOF. The case $0 < p < q < \infty$ is the standard Maurey theorem and we refer the reader to [Ma] or [W] for this. We shall only consider the case where $p < 0$.

One direction of the theorem immediately follows from (6.16). In fact, if we have (6.18), then, for every $g = (\sum_{k=1}^{m} c_k |f_k|^q)^{1/q} \in \tilde{V}_q$, where $\sum_{k=1}^{m} c_k = 1$, $f_k \in V$,

$$\|g\|_{p,d\mu}^q = \|gv^{-\frac{1}{p}}\|_{p,vd\mu}^q \leq \|gv^{-\frac{1}{p}}\|_{q,vd\mu}^q = \sum_{k=1}^{m} c_k \|f_k v^{-\frac{1}{p}}\|_{q,vd\mu}^q \leq \sigma^q.$$

We turn to the proof of sufficiency. We can assume that

$$\sigma = \sup\{\|g\|_{p,d\mu} : \ g \in \tilde{V}_q\}.$$

First we prove that if (g_n) is a sequence in $\tilde{V}_q$ such that $g_n \to g$ μ-a.e., where g is an extended-valued measurable function $g : \Omega \to [0, \infty)$, then $\limsup_{n\to\infty} \|g_n\|_{p,d\mu} \leq \|g\|_{p,d\mu} \leq \sigma$.

Notice that from Fatou's lemma

$$\|g\|_{p,d\mu} \geq \limsup_{n\to\infty} \|g_n\|_{p,d\mu},$$

and so we only need to show that $\|g\|_{p,d\mu} \leq \sigma$. Suppose $t > 0$ and let $g_n' = ((1-t)g_n^q + th^q)^{\frac{1}{q}} \in \tilde{V}_q$. Then g_n' converges a.e. to $((1-t)g^q + th^q)^{\frac{1}{q}}$. Now if $p < 0$, it follows easily from the dominated convergence theorem (note that $h^p \in L_1(d\mu)$ because $\|h\|_{p,d\mu} > 0$) that

$$\sigma \geq \lim_{n\to\infty} \|g_n'\|_{p,d\mu} = \|((1-t)g^q + th^q)^{\frac{1}{q}}\|_{p,d\mu}. \tag{6.20}$$

By Lemma 6.21,

$$\|((1-t)g^q + th^q)^{\frac{1}{q}}\|_{p,d\mu} \geq ((1-t)\|g\|_{p,d\mu}^q + t\|h\|_{p,d\mu}^q)^{\frac{1}{q}},$$

so

$$((1-t)\|g\|_{p,d\mu}^q + t\|h\|_{p,d\mu}^q)^{\frac{1}{q}} \leq \sigma.$$

Since $t > 0$ is arbitrary, we get $\|g\|_{p,d\mu} \leq \sigma$.

Take any sequence (g_n) in $\tilde{V}_q$ such that $\lim_{n\to\infty} \|g_n\|_{p,d\mu} = \sigma$. We can use Lemma 6.23 to find a sequence g_n' so that $(g_n')^q \in \mathrm{co}\{g_k^q : k \geq n\}$ and g_n' converges a.e to some g. Then by Lemma 6.21 we have that $\|g_n'\|_{p,d\mu} \to \sigma$ and the preceding argument shows that $\|g\|_{p,d\mu} = \sigma$.

Now if $f \in V$, we have that $((1-t)g^q + t|f|^q)^{\frac{1}{q}} \in \tilde{V}_q$ for any $t > 0$, so

$$\|((1-t)g^q + t|f|^q)^{\frac{1}{q}}\|_{p,d\mu} \leq \sigma = \|g\|_{p,d\mu}.$$

This implies

$$\int ((1-t)g^q + t|f|^q)^{\frac{p}{q}} d\mu \geq \int g^p d\mu.$$

Let $F_M = \min(|f|g^{-1}, M)$. Then

$$\int g^p((1 + t(F_M^q - 1))^{\frac{p}{q}} - 1) d\mu \geq 0.$$

Next we divide both sides of the above inequality by t:

$$\int g^p \frac{(1 + t(F_M^q - 1))^{\frac{p}{q}} - 1}{t} d\mu \geq 0.$$

It follows from the dominated convergence theorem, since $g^p \in L_1(d\mu)$, that (as $t \to 0$)

$$\int g^p (F_M^q - 1) d\mu \leq 0.$$

Now, by the monotone convergence theorem (as $M \to \infty$)

$$\int g^{p-q}|f|^q d\mu \leq \int g^p d\mu = \sigma^p.$$

Let $v = \sigma^{-p} g^p$. Then

$$\|v^{-\frac{1}{p}} f\|_{q,v d\mu} \leq \sigma.$$

□

We are now ready to prove Theorem 6.18. We start with the case where $-1 < p < 1$, $p \neq 0$. In this case the definition of embedding in L_p can be written in the form of the Levy representation: a space $X = (\mathbb{R}^n, \|\cdot\|_X)$ embeds in L_p if and only if there exists a finite Borel measure μ on the sphere such that, for every $f \in \mathbb{R}^n$, $f \neq 0$,

$$\|f\|_X^p = \int_{S^{n-1}} |(f,u)|^p d\mu(u). \tag{6.21}$$

It is more convenient for us to use the letter f for elements of X, so $f \in \mathbb{R}^n$. This notation may be a little confusing but relates to the notation in the original factorization theorem of Maurey and Nikishin. Recall that the embedding $U : X \to \mathcal{M}(S^{n-1}, \mu)$ is defined by $Uf = (f, u)$, $u \in S^{n-1}$. Then by (6.21), for every $f \in X$, we have

$$\|f\|_X = \|Uf\|_{p,d\mu}. \tag{6.22}$$

THEOREM 6.25. *Suppose $-1 < p < q < 1$ with $p \neq 0$ and $q > 0$. Suppose that X is an n-dimensional normed space that embeds in L_p with the corresponding measure μ. Then there exist a function $v \in L_1(S^{n-1}, \mu)$ with $v \geq 0$ a.e., $\int_{S^{n-1}} v\, d\mu = 1$ and an n-dimensional subspace Y of $L_q(S^{n-1}, vd\mu)$ such that*

$$d(X, Y) \leq \frac{(\cos(\frac{p\pi}{2}))^{\frac{1}{p}}}{(\cos(\frac{q\pi}{2}))^{\frac{1}{q}}} = c(p, q).$$

PROOF. We can assume that μ is a probability measure on S^{n-1} because the Banach-Mazur distance from X to any normed space does not change if we introduce a new norm in X, which differs from the old one by a constant multiple.

We are going to find a function v such that, for every $f \in X$,

$$\|f\|_X \leq \|v^{-\frac{1}{p}} Uf\|_{q,vd\mu} \leq c(p, q)\|f\|_X. \tag{6.23}$$

The latter would mean that the operator $Tx = v^{-\frac{1}{p}}Ux$ acting from X onto a subspace Y of $L_q(S^{n-1}, vd\mu)$ satisfies $\|T\|\|T^{-1}\| \leq c(p, q)$, which would prove our claim.

We first note that the lower bound in (6.23) holds for any non-negative integrable function v on S^{n-1} satisfying $\int_{S^{n-1}} vd\mu = 1$. In fact, since $vd\mu$ is a probability measure on the sphere, by (6.22) and (6.16),

$$\|f\|_X = \|Uf\|_{p,d\mu} = \|v^{-\frac{1}{p}} Uf\|_{p,vd\mu} \leq \|v^{-\frac{1}{p}} Uf\|_{q,vd\mu}.$$

We now use Theorem 6.24 to choose a specific function v providing the upper bound in (6.23). Let

$$V = \{Uf : \ f \in X, \|f\|_X = 1\}.$$

Consider $g \in \tilde{V}_q$ such that $g = (\sum_{k=1}^m c_k |Uf_k|^q)^{\frac{1}{q}}$. Here $\sum_{k=1}^m c_k = 1$, $f_1, ..., f_m \in X$, $\|f_k\|_X = 1$, $k = 1, ..., m$.

Let $(\eta_k)_{k=1}^m$ be independent normalized symmetric q-stable random variables. Then, for any $f_1, ..., f_m \in X$,

$$(\mathbb{E}\|\sum_{k=1}^m \eta_k c_k^{1/q} Uf_k\|_{p,d\mu}^p)^{\frac{1}{p}} = \left(\int_{S^{n-1}} \mathbb{E}|\sum_{k=1}^m \eta_k c_k^{1/q} Uf_k(u)|^p d\mu(u)\right)^{\frac{1}{p}}. \tag{6.24}$$

Next, by Lemma 6.3, for every $u \in S^{n-1}$, the random variables

$$\sum_{k=1}^m \eta_k c_k^{1/q} Uf_k(u) \qquad \text{and} \qquad (\sum_{k=1}^m c_k |Uf_k(u)|^q)^{\frac{1}{q}} \eta_1$$

have equal distributions, so, also using (2.57), we get that the right-hand side of (6.24)

$$= (\mathbb{E}|\eta_1|^p)^{\frac{1}{p}} \|(\sum_{k=1}^m c_k |Uf_k|^q)^{\frac{1}{q}}\|_{p,d\mu}$$

$$= \left((\cos(p\pi/2))^{-1} \frac{\Gamma(\frac{-p}{q})}{q\Gamma(-p)}\right)^{\frac{1}{p}} \|(\sum_{k=1}^m c_k |Uf_k|^q)^{\frac{1}{q}}\|_{p,d\mu}.$$

Hence,

$$\|g\|_{p,d\mu} = \|(\sum_{k=1}^{m} c_k |Uf_k|^q)^{\frac{1}{q}}\|_{p,d\mu}$$

$$= \left((\cos(\frac{p\pi}{2}))^{-1}\frac{\Gamma(\frac{-p}{q})}{q\Gamma(-p)}\right)^{-\frac{1}{p}} (\mathbb{E}\|\sum_{k=1}^{m}\eta_k c_k^{1/q} Uf_k\|_{p,d\mu}^{p})^{\frac{1}{p}}.$$

Recall that $f_1, \ldots, f_m \in X$, where X is a Banach space, so by the triangle inequality and (6.22),

$$\|\sum_{k=1}^{m}\eta_k c_k^{1/q} Uf_k\|_{p,d\mu} = \|\sum_{k=1}^{m}\eta_k c_k^{1/q} f_k\|_X$$

$$\leq \sum_{k=1}^{m}|\eta_k| c_k^{1/q}\|f_k\|_X = \sum_{k=1}^{m}|\eta_k| c_k^{1/q},$$

where the last equality follows from the fact that every Uf_k belongs to V, so $\|f_k\|_X = 1$. Finally,

$$\|g\|_{p,d\mu} \leq \left((\cos(\frac{p\pi}{2}))^{-1}\frac{\Gamma(\frac{-p}{q})}{q\Gamma(-p)}\right)^{-\frac{1}{p}} (\mathbb{E}(\sum_{k=1}^{m}|\eta_k| c_k^{1/q})^{p})^{\frac{1}{p}} \leq \frac{(\cos(\frac{p\pi}{2}))^{\frac{1}{p}}}{(\cos(\frac{q\pi}{2}))^{\frac{1}{q}}}$$

by Lemma 6.20 and the fact that $\sum_{k=1}^{m} c_k = 1$. Thus for every $g \in \tilde{V}_q$, $\|g\|_{p,d\mu} \leq c(p,q)$. Now the result follows from Theorem 6.24, namely, there exists v with the required properties so that $\|v^{-\frac{1}{p}}Uf\|_{q,vd\mu} \leq c(p,q)$ for each $f \in X$ with $\|f\|_X = 1$. By homogeneity, this implies the upper bound in (6.23). □

This proof does not work for $p \leq -1$ for the reason that the embedding in L_p is no longer given by (6.21). Because of that we cannot write the norm directly by (6.22). For the upper bound in (6.23) we can get around it using Lemma 6.22 and Lemma 6.20.

However, the lower bound in (6.23), where we can no longer use monotonicity of L_p-norms, becomes more difficult, and we apply several results from the Banach space theory. We still try to use Lemma 6.22, but now we have to get a lower bound for the expectation in the left-hand side. To do this, we need the number m of vectors f_k to be greater than $-p$. The key argument in the proof is inequality (6.30), which allows us to get the lower estimate (6.28) for a subspace of small (comparable to p) codimension. We then simply embed the remaining parts of X into Euclidean spaces of dimension comparable to p with the Banach-Mazur distance depending on p only. This way we construct an embedding of X into the q-sum of an L_q-space and a Euclidean space. Such a sum is itself a subspace of a larger L_q-space, since Euclidean spaces embed isometrically in L_q.

THEOREM 6.26. *If $-\infty < p < q < 1$, $p \leq -1$ and $q > 0$, then there exists a constant $C = C(p,q)$ so that, for any n with $-n < p$, if X is an n-dimensional normed space that embeds into L_p, then there is a subspace Y of L_q so that $d(X,Y) \leq C$.*

PROOF. Let $\nu = [-p]$. We have $\nu \geq 1$. Let μ be the measure on the sphere S^{n-1} corresponding to the space X by the definition of embedding in L_p; see Definition 6.14. As before, our measure space will be (S^{n-1}, μ), and we assume that μ is a probability measure. The operator $U : X \to \mathcal{M}(S^{n-1}, \mu)$ is also the same as before.

By Lemma 6.22, for any $f_1, ..., f_m \in X$, whose linear span contains X,

$$(\mathbb{E}\|\sum_{k=1}^{m} \eta_k f_k\|_X^p)^{\frac{1}{p}} = c(p,q)\|(\sum_{k=1}^{m} |Uf_k|^q)^{\frac{1}{q}}\|_{p,d\mu}, \tag{6.25}$$

whenever $(\eta_k)_{k=1}^m$ are independent normalized symmetric q-stable random variables. Hence, estimating the left-hand side of (6.25) by the triangle inequality in the normed space X and by Lemma 6.20, we have

$$\|(\sum_{k=1}^{m} |Uf_k|^q)^{\frac{1}{q}}\|_{p,d\mu} \le C(p,q)(\sum_{k=1}^{m} \|f_k\|_X^q)^{\frac{1}{q}}, \tag{6.26}$$

where $C(p,q)$ is the product of constants appearing in Lemma 6.20 and Lemma 6.22. Let $V = \{Uf : f \in X, \|f\|_X = 1\}$. Using inequality (6.26), one can immediately verify the conditions of Theorem 6.24. The existence of a function h required in Theorem 6.24 follows, for example, from (6.30) (see below) and Lemma 6.22. Therefore, there exists $v \in L_1$ with $v \ge 0$ and $\int v\, d\mu = 1$ so that for all $f \in X$ we have

$$\|v^{-\frac{1}{p}} Uf\|_{q,vd\mu} \le C\|f\|_X, \tag{6.27}$$

with $C = C(p,q)$ depending on p, q only.

As already mentioned, for $p \le -1$ the lower bound $\|v^{-\frac{1}{p}} Uf\|_{q,vd\mu} \ge c\|f\|_X$ no longer follows from monotonicity of L_p-norms, and we need to adjust the embedding.

We first note that, since $\nu + 1 > -p$, the expectation

$$\mathbb{E}(\max_{1\le k\le \nu+1} |\eta_k|)^p$$

is finite and depends only on p and q. We put

$$c_0 = c_0(p,q) = (\mathbb{E}(\max_{1\le k\le \nu+1} |\eta_k|)^p)^{\frac{1}{p}} > 0.$$

Consider a number c such that $0 < c < c_0(\nu+1)^{-\frac{1}{q}}(c(p,q))^{-1}$, where $c(p,q)$ is the constant from Lemma 6.22. Let us prove that there is a subspace E of X of dimension at most ν and a subspace F of X of codimension at most ν so that if $f \in F$, then

$$\|v^{-\frac{1}{p}} Uf\|_{q,vd\mu} \ge c\, d(f,E), \tag{6.28}$$

where $d(f,E) := \inf\{\|f - e\|_X : e \in E\}$. To see this, suppose the contrary. Then, by induction we can pick $f_1, \dots, f_{\nu+1} \in X$ and $\phi_1, \dots, \phi_{\nu+1} \in X^*$ such that $\|f_k\|_X = \|\phi_k\|_{X^*} = 1$ for $k = 1, 2, \dots, \nu+1$, $\phi_k(f_j) = 0$ if $j \ne k$ and $\phi_k(f_k) = 1$ so that $d(f_k, [f_1, \dots, f_{k-1}]) = 1$, but for each k

$$\|v^{-\frac{1}{p}} Uf_k\|_{q,vd\mu} < c. \tag{6.29}$$

Here $[f_1, \dots, f_{k-1}]$ stands for the linear span. Now

$$\|\sum_{k=1}^{\nu+1} \eta_k f_k\|_X \ge \max_{1\le i\le \nu+1} \left|\phi_i\left(\sum_{k=1}^{\nu+1} \eta_k f_k\right)\right| = \max_{1\le i\le \nu+1} |\eta_k|$$

and

$$(\mathbb{E}\|\sum_{k=1}^{\nu+1} \eta_k f_k\|_X^p)^{\frac{1}{p}} \ge \left(\mathbb{E} \max_{1\le i\le \nu+1} |\eta_i|^p\right)^{\frac{1}{p}} = c_0. \tag{6.30}$$

But we also have an upper bound: by Lemma 6.22 and (6.29),

$$(\mathbb{E}\|\sum_{k=1}^{\nu+1}\eta_k f_k\|_X^p)^{\frac{1}{p}} = c(p,q)\|(\sum_{k=1}^{\nu+1}|Uf_k|^q)^{\frac{1}{q}}\|_{p,d\mu}$$

$$= c(p,q)\|(\sum_{k=1}^{\nu+1}|v^{-1/p}Uf_k|^q)^{\frac{1}{q}}\|_{p,vd\mu} \leq c(p,q)\|(\sum_{k=1}^{\nu+1}|v^{-1/p}Uf_k|^q)^{1/q}\|_{q,vd\mu}$$

$$= c(p,q)\left(\sum_{k=1}^{\nu+1}\|v^{-1/p}Uf_k\|_{q,vd\mu}^q\right)^{1/q} \leq c(\nu+1)^{\frac{1}{q}}c(p,q),$$

so $c_0 \leq c(\nu+1)^{\frac{1}{q}}c(p,q)$, which gives a contradiction.

Next we use John's theorem, [J], saying that the Banach-Mazur distance from any m-dimensional normed space E to the m-dimensional Euclidean space is at most $\sqrt{m}$; see [Ga3, Th.4.2.12] or [MiS, p.10] for a simple proof. There exists a bounded linear operator $S_0 : E \to \ell_2^{\nu+1}$, $m = \dim E \leq \nu$ with $|S_0 e|_2 \geq \|e\|_X$ for $e \in E$ and $\pi_2(S_0) \leq \sqrt{\nu}$, where $\pi_2(S_0)$ denotes the 2-summing norm of the operator S_0 (see [Pi2, p. 33]). Moreover S_0 can be extended to a bounded operator $S : X \to \ell_2^m$ with $\|S\| \leq \pi_2(S) \leq \sqrt{\nu}$ ([Pi2, Corollary 3.9]).

There is also a linear operator $T : X \to \ell_2^m$ with $\|T\| \leq \sqrt{\nu}$ and $|Tf|_2 \geq d(f,F)$ for $f \in X$. In fact, again by John's theorem, there exists an operator $T' : X/F \to l_2^m$ with $\|T'\| \leq \sqrt{m}$ and $\|T'\| \geq 1$. Then we can put $T = T'Q$ where $Q : X \to X/F$ is the quotient map.

Consider the following q-sum of spaces $Y = (\ell_2^m \oplus \ell_2^m \oplus L_q(vd\mu))_q$ which embeds isometrically into L_q, as the q-sum of spaces embedding isometrically in L_q (the space l_2^m embeds in L_q by Corollary 6.7). We define an operator $D : X \to Y$ by $Df = (Sf, Tf, Jf)$ where $Jf = v^{-\frac{1}{p}}Uf$. Then, by (6.27) and estimates for the norms of S and T, $\|D\| \leq C_1 = C_1(p,q)$, and our goal is to find a lower bound. Assume $\|f\|_X = 1$; we show that $\|Df\| \geq \delta$ where $\delta > 0$ is chosen so that

$$\left[1 + (\sqrt{\nu}+1)((2C^q+1)^{\frac{1}{q}}c^{-1}+2)\right] < \delta^{-1},$$

when C is defined by inequality (6.27). We may pick $g \in F$ with $\|f-g\|_X < 2d(f,F)$. Then we can find $h \in E$ so that $\|g-h\|_X < 2d(g,E)$. Assume $\|Df\|_q < \delta$. Then $|Tf|_2 < \delta$ (also both numbers $|Sf|_2$ and $\|Jf\|_{q,vd\mu}$ are less than δ) so that $d(f,F) < \delta$. Since $|Tf|_2 \geq d(f,F) > \|f-g\|_X$, we have $\|g\|_X > 1-2\delta$. Now, from inequality (6.27) and since $0 < q < 1$, we get

$$\|Jg\|_{q,\nu d\mu}^q < \|Jf\|_{q,\nu d\mu}^q + \|Jf - Jg\|_{q,\nu d\mu}^q$$

$$\leq \|Jf\|_{q,\nu d\mu}^q + 2C^q\delta^q < (2C^q+1)\delta^q$$

where the first inequality is a simple property of L_q-norms with $0 < q < 1$; see [KPR] for the theory of these spaces.

Now, inequality (6.28) gives $d(g,E) < (2C^q+1)^{\frac{1}{q}}c^{-1}\delta$. Hence

$$\|f-h\|_X < ((2C^q+1)^{\frac{1}{q}}c^{-1}+2)\delta.$$

Since $\|S\| \leq \sqrt{\nu}$ and $|Sf|_2 < \delta$,

$$|Sh|_2 \leq |Sf|_2 + |S(f-h)|_2 \leq \left[1 + \sqrt{\nu}((2C^q+1)^{\frac{1}{q}}c^{-1}+2)\right]\delta.$$

However

$$|Sh|_2 \geq \|h\|_X \geq \|f\|_X - \|f-h\|_X \geq 1 - ((2C^q+1)^{\frac{1}{q}}c^{-1}+2)\delta.$$

This gives a contradiction. Therefore, we have $\|Df\| \geq \delta$ for every $f \in X$, $\|f\|_X = 1$, where δ depends only on p and q. This completes the proof. □

6.4. Remarks and further results

In the first two sections of Chapter 6 we explored the connections between positive definite functions, stable measures and embeddings of normed spaces in L_p.

In the 1920's and early 1930's, P. Levy [Le] proved that, for any finite dimensional subspace $(\mathbb{R}^n, \|\cdot\|)$ of L_p with $0 < p \leq 2$, the function $\exp(-\|\cdot\|^p)$ is positive definite, and the probability measure μ on $\mathbb{R}^n$, whose Fourier transform is equal to this function, has a remarkable property: if $X = (X_1, ..., X_n)$ is a random vector with distribution μ, then every linear combination of coordinates $\sum_{i=1}^n a_i X_i$ has the same distribution as $\|a\|X_1$. This discovery of P. Levy gave a start to the theory of stable processes that has many applications to different areas of mathematics and science and still remains under intensive development; see, for example, [Lin].

Generalizing Levy's results, Eaton [Ea] introduced the general concept of a K-isotropic measure and showed that the problem of finding all such measures is equivalent to characterizing the class $\Phi(K)$ of functions $f : \mathbb{R} \to \mathbb{R}$ for which $f(\|\cdot\|_K)$ is a positive definite function on $\mathbb{R}^n$. Our Lemma 6.1 is the result of Eaton. If E is a normed space with the unit ball B_E, we write $\Phi(E) = \Phi(B_E)$.

The classes $\Phi(K)$ have later been studied by a number of authors. Schoenberg [Sc2] proved that $f \in \Phi(B_2^n)$ if and only if

$$f(t) = \int_0^\infty \Omega_n(tr)\, d\lambda(r)$$

where $\Omega_n(|\cdot|_2)$ is the Fourier transform of the uniform probability measure on the sphere S^{n-1} and λ is a finite measure on $[0, \infty)$. In the same paper, Schoenberg proved an infinite dimensional version of this result: $f \in \Phi(\ell_2)$ if and only if

$$f(t) = \int_0^\infty \exp(-t^2 r^2)\, d\lambda(r).$$

Bretagnolle, Dacunha-Castelle and Krivine [BDK] proved a similar result for the classes $\Phi(\ell_q)$ for all $q \in (0,2)$ (one just has to replace 2 by q in the formula) and proved that for $q > 2$ the classes $\Phi(\ell_q)$ (corresponding to infinite dimensional ℓ_q-spaces) are trivial, i.e. contain constant functions only. Cambanis, Keener and Simons [CKS] obtained a similar representation for the classes $\Phi(B_1^n)$. Richards [Ri] partially characterized the classes $\Phi(B_q^n)$ for $0 < q < 2$. Aharoni, Maurey and Mityagin [AMM] proved that if E is an infinite dimensional Banach space with a symmetric basis $\{e_n\}_{n=1}^\infty$ such that $\lim\limits_{n\to\infty} \|e_1 + \cdots + e_n\|/n^{1/2} = 0$, then the class $\Phi(E)$ is trivial. For other related results see [As], [BR], [CR], [Kue], [Kur], [KuS], [MisR], [Gn1], [Gn2], [Za3]. One can find more detailed information on isotropic measures in the survey [MisS] and in the book by Misiewicz [Mis3]. In all the results mentioned above the classes $\Phi(E)$ appear to be non-trivial only if E is a subspace of L_q with $0 < q \leq 2$. It is still not known if there exist other spaces E for which the class $\Phi(E)$ is non-trivial.

In 1938, Schoenberg [Sc1] posed the problem of finding the exponents $0 < p < 2$ for which the function $\exp(-\|\cdot\|_q^p)$ is positive definite, where the norm is taken from one of the spaces ℓ_q^n with $2 < q \le \infty$. For the spaces ℓ_∞^n, the problem was solved in 1989 by Misiewicz [Mi2], and, for the spaces ℓ_q^n with $2 < q < \infty$, the answer was given in [K1] in 1991 (note that, for $1 \le p \le 2$, Schoenberg's question was answered earlier by Dor [Do]). The answers turned out to be the same for both classes of spaces: the function $\exp(-\|\cdot\|_q^p)$ is not positive definite if the dimension of the space is greater than 2, and for $n = 2$ the function is positive definite if and only if $0 < p \le 1$. Moreover, Misiewicz was able to prove that for $n \ge 3$ the classes $\Phi(\ell_\infty^n)$ are trivial. The result from [K1] was not that strong, but, after that paper appeared, Lisitsky [Lis] and Zastavnyi [Za1], [Za2] proved that, for $n \ge 3$ and $q > 2$, the classes $\Phi(\ell_q^n)$ are also trivial. In Section 6.2, we presented a result from [KL] generalizing the solution of Schoenberg's problem in the case $2 < q < \infty$; see Theorem 6.11. The result of Misiewicz proving Schoenberg's conjecture for $q = \infty$ is presented in Theorem 6.13 with a proof from [K4]. For other related results see [DK].

One of the reasons for Schoenberg to ask his questions was his expectation that positive definiteness of certain functions can be applied to isometric embeddings of Banach spaces. The connection between positive definite functions and isometric embeddings into L_p was discovered by P. Levy, who essentially proved Theorem 6.6. It is possible to prove this result in a little stronger form, which was done by Misiewicz [Mis1] and Kuritsyn [Kur]; see Lemma 6.5. The infinite dimensional version of Theorem 6.6 was established by Bretagnolle, Dacunha-Castelle, and Krivine [BDK], who also proved Corollary 6.7. Another characterization of subspaces of L_p in terms of the Fourier transform, which works for all $p > 0$ that are not even integers, was given in [K2]; see Theorem 6.10.

The motivation and contents of Section 6.3 were presented in detail in the Introduction. Let us mention only that the concept of embedding of a normed space in L_p with $p < 0$ was first introduced in [K12] and then studied in more detail in [K14]. Theorem 6.15 was proved in [K14], Theorem 6.17 in [K10]. Note that generalized k-intersection bodies of Zhang (see Section 4.5) also admit a functional analytic interpretation as subspaces of L_p-spaces of vector-valued functions with $p < 0$; see [K14].

Theorem 6.18 was proved in [KK2]. The constant $c(k, q)$ in Theorem 6.18 tends to infinity as $q \to 1$, so the method of this paper does not work in the case $q = 1$. Since the unit balls of subspaces of L_1 are polar projection bodies (see Theorem 8.6), the statement of Theorem 6.18 in the case $q = 1$ would mean, if true, that intersection bodies are isomorphically equivalent to polar projection bodies. This would represent a big step towards resolving one of the mysteries of convex geometry – the duality between sections and projections. The corresponding question in the theory of L_p-spaces is also open and is the matter of Kwapien's problem [Kw] posed in 1970: is every Banach subspace of L_p with $0 < p < 1$ isomorphic to a subspace of L_1? For partial results on Kwapien's problem see [Ka]. Note that the isometric version of Kwapien's problem has a negative answer; see [K5] and [KK1]. This means, in particular, that not every intersection body is a polar projection body. The result of [KK1] also implies that there exists an intersection body which is not the unit ball of any subspace of L_q with $q \in (0, 1)$ (isometrically).

The connection between convex geometry and the theory of L_p-spaces provides a continuous path from polar projection bodies to intersection bodies, going from the space L_1 to L_{-1}. The geometry of subspaces of L_1 and projection bodies is related to the classical Brunn-Minkowski theory, because it corresponds to the Minkowski summation. On the other hand, the geometry of L_{-1} and intersection bodies corresponds to the radial summation and is the subject of the dual Brunn-Minkowski theory, introduced by Lutwak [Lu1]. It would be interesting to extend the Brunn-Minkowski theory to the intermediate spaces L_p, $p \in (-1, 1)$. Note that the extension of the Brunn-Minkowski theory corresponding to L_p-spaces with $p > 1$, the so-called Brunn-Minkowski-Firey theory, was developed by Lutwak [Lu2], [Lu3] with many results analogous to basic facts from the classical theory. Many more interesting results in this direction can be found in a series of papers by Lutwak, Yang and Zhang; see [LYZ1]–[LYZ4].

The connection between classes of convex bodies that are the unit balls of subspaces of L_p and different kinds of summation can be interpreted in a purely geometric way. Suppose that we start with the Euclidean ball and are allowed to construct new bodies using three operations: linear transformations, closure in the radial metric and p-summation defined for $p \neq 0$ by

$$\|x\|_K^p = \|x\|_{K_1}^p + \|x\|_{K_2}^p$$

and for $p = 0$ by

$$\|x\|_K = \sqrt{\|x\|_{K_1} \|x\|_{K_2}}.$$

What classes of bodies do we get? It is well known that in the case $p = 1$ we get all unit balls of subspaces of L_1 (or polar projection bodies; see [Ga3, Corollary 4.1.12]), while in the case $p = -1$ we get all intersection bodies (or, in other words, the unit balls of spaces that embed in L_{-1}), as proved by Goodey and Weil [GW2]. In particular, in the case $p = 1$ we get all origin-symmetric convex bodies in $\mathbb{R}^2$, but not in $\mathbb{R}^3$, and in the case $p = -1$ we get all origin-symmetric convex bodies in $\mathbb{R}^4$, but not in $\mathbb{R}^5$. As proved in [KKYY], if $p = 0$, we get all origin-symmetric convex bodies in dimension 3, but not in dimension 4 (see Example 1 in Section 6.3, where one can also find the definition of embedding in L_0 introduced in [KKYY]).

There are other results where Fourier methods are applied to the study of subspaces of L_p. For example, spherical harmonics were used by Schneider in the paper [Sch2] to prove that there exist nontrivial zonoids whose polars are also zonoids. This means that there exists a non-Hilbertian subspace of L_1 whose dual space is also a subspace of L_1 (isometrically). A related problem, due to Lindenstrauss, is whether the sequence α_n converges to 1, where α_n is the supremum of $d(E, l_2^n)$ over all n-dimensional normed spaces E such that both E and E^* embed isometrically in L_1. Since the examples in Schneider's paper are small smooth perturbations of the Euclidean ball, it would be interesting to find examples of non-smooth zonoids whose polars are zonoids. In dimensions 3 and 4 this was done by Lonke [Lo]. The paper [K3] modifies Schneider's method to give examples of non-Hilbertian normed spaces that, at the same time, embed in L_p with $p < 2$ and in L_q with $q > 2$.

CHAPTER 7

Extremal Sections of ℓ_q-Balls

A very concrete problem of finding the extremal sections of ℓ_q-balls has motivated some beautiful mathematics. The problem seems to be quite difficult, for example the maximal hyperplane sections of ℓ_q-balls with $2 < q < \infty$ are still unknown. In this chapter we prove those results that involve Fourier techniques and discuss other results in the last section.

7.1. The case of the cube, K. Ball's theorem

We start with a remarkable result of K. Ball on the maximal section of the cube. Let $Q_n = \frac{1}{2}B_\infty^n$ be the unit cube in $\mathbb{R}^n$.

THEOREM 7.1. *For every* $\xi \in S^{n-1}$,

$$\mathrm{Vol}_{n-1}(Q_n \cap \xi^\perp) \leq \sqrt{2}$$

with equality for $\xi = (\frac{1}{\sqrt{2}}, \frac{1}{\sqrt{2}}, 0, ..., 0)$.

PROOF. By symmetry and continuity, we can assume that $\xi_k > 0$ for every k. Suppose first that there exists k so that $\xi_k > 1/\sqrt{2}$. Project the section $Q_n \cap \xi^\perp$ to the coordinate hyperplane $x_k = 0$ in $\mathbb{R}^n$. The $(n-1)$-dimensional volume of the projection is less than or equal to 1, because this projection is contained in the projection of the whole cube. On the other hand, the $(n-1)$-dimensional volume of the projection is equal to the area of the section times the cosine of the angle between this section and the plane $x_k = 0$. This cosine is equal to ξ_k, so

$$\xi_k \mathrm{Vol}_{n-1}(Q_n \cap \xi^\perp) \leq 1,$$

and the result follows from the assumption on ξ_k.

Suppose now that each $\xi_k \leq 1/\sqrt{2}$. By Theorem 3.1 with $t = 0$ (note that 2^{n-1} disappears when we consider Q_n in place of B_∞^n),

$$\mathrm{Vol}_{n-1}(Q_n \cap \xi^\perp) \leq \int_{\mathbb{R}} \prod_{k=1}^n \left| \frac{\sin(\pi t \xi_k)}{\pi t \xi_k} \right| dt.$$

By Hölder's inequality (recall that $\xi_1^2 + ... + \xi_n^2 = 1$), the quantity in the right-hand side is less than or equal to

$$\prod_{k=1}^n \left(\int_{\mathbb{R}} \left| \frac{\sin(\pi t \xi_k)}{\pi t \xi_k} \right|^{1/\xi_k^2} dt \right)^{\xi_k^2}$$

$$= \prod_{k=1}^n \left(\frac{1}{\xi_k} \right)^{\xi_k^2} \left(\int_{\mathbb{R}} \left| \frac{\sin(\pi z)}{\pi z} \right|^{1/\xi_k^2} dz \right)^{\xi_k^2}.$$

By Ball's inequality (see Lemma 7.4) with $t = 1/\xi_k^2 \geq 2$, the latter is less than or equal to

$$\prod_{k=1}^{n} \left(\frac{1}{\xi_k}\right)^{\xi_k^2} \left(\sqrt{2\xi_k^2}\right)^{\xi_k^2} = \sqrt{2}.$$

□

In the next three lemmas we prove Ball's inequality used in the last step of the proof of Theorem 7.1. We follow the proof given by Nazarov and Podkorytov in [NP].

Let (X, μ) be a measure space and $f : X \to \mathbb{R}$ a measurable function. The distribution function F of f is defined by

$$F(y) = \mu\{x \in X : \ f(x) > y\},$$

for every $y \in \mathbb{R}$.

LEMMA 7.2. *Let f, g be measurable functions on (X, μ), and denote by F, G their distribution functions. Assume that F, G are finite for every $y \in \mathbb{R}$ and that $f - g$ is an integrable function on X. Then*

$$\int_X (f(x) - g(x))\, d\mu(x) = \int_{\mathbb{R}} (F(y) - G(y))\, dy.$$

PROOF. Let $h = \min(f, g)$ and let H be the distribution function of h. Since $h \leq f$, by the Fubini theorem,

$$\int_X (f(x) - h(x)) d\mu(x) = \int_X \left(\int_{h(x)}^{f(x)} dy \right) d\mu(x)$$

$$= \int_{\mathbb{R}} \mu\{x : \ h(x) \leq y < f(x)\}\, dy = \int_{\mathbb{R}} (F(y) - H(y))\, dy.$$

A similar equality holds for g and h. The result follows by subtraction. □

LEMMA 7.3. *Suppose that, in addition to the assumptions of Lemma 7.2, f, g are non-negative functions and there exists $y_0 \in [0, \infty)$ so that $F(y) \leq G(y)$ on $[0, y_0]$ and $F(y) \geq G(y)$ on $[y_0, \infty)$. Suppose that, for some $0 < s < t$, the functions $f^s - g^s$ and $f^t - g^t$ are integrable on X. Then*

$$\frac{1}{s y_0^s} \int_X (f^s - g^s)\, d\mu \leq \frac{1}{t y_0^t} \int_X (f^t - g^t)\, d\mu.$$

PROOF. For $s > 0$, the distribution function of f^s is equal to $F(y^{1/s})$. By Lemma 7.2,

$$\phi(s) = \frac{1}{s y_0^s} \int_X (f^s - g^s)\, d\mu = \frac{1}{s y_0^s} \int_0^\infty (F(y^{1/s}) - G(y^{1/s}))\, dy$$

$$= \frac{1}{y_0^s} \int_0^\infty y^{s-1} (F(y) - G(y))\, dy.$$

Therefore,

$$\phi(t) - \phi(s) = \frac{1}{y_0} \int_0^\infty \left(\left(\frac{y}{y_0}\right)^{t-1} - \left(\frac{y}{y_0}\right)^{s-1} \right) (F(y) - G(y))\, dy,$$

and the integrand is always non-negative by our assumptions on F and G. □

We now prove Ball's inequality.

LEMMA 7.4. *(Ball's Integral Inequality) For $t \geq 2$,*

$$\int_{-\infty}^{\infty} \left|\frac{\sin(\pi x)}{\pi x}\right|^t dx \leq \sqrt{\frac{2}{t}}. \tag{7.1}$$

PROOF. We apply Lemma 7.3 to the situation where (X, μ) is $[0, \infty)$ with Lebesgue measure,

$$g(x) = |\sin(\pi x)/(\pi x)| \qquad \text{and} \qquad f(x) = \exp(-\pi x^2/2).$$

It is easy to check that $\int_0^\infty g^2 = \int_0^\infty f^2$ (one can use the formula from Theorem 3.1 to compute the first integral). Also, $\int_0^\infty f^t = \sqrt{2/t}$. Let F and G be the distribution functions of f and g, respectively. The only thing that remains to be proved is that the distribution functions F and G satisfy the condition of Lemma 7.3.

Since $0 < f, g < 1$, we have $F = G = 0$ on $[1, \infty)$, and it is enough to consider F and G on the interval $(0, 1)$ only.

Consider the numbers $y_m = \max_{[m,m+1]} g$, $m \in \mathbb{N}$. Clearly,

$$\frac{1}{\pi(m + 1/2)} = g(m + 1/2) < y_m < \frac{1}{\pi m}.$$

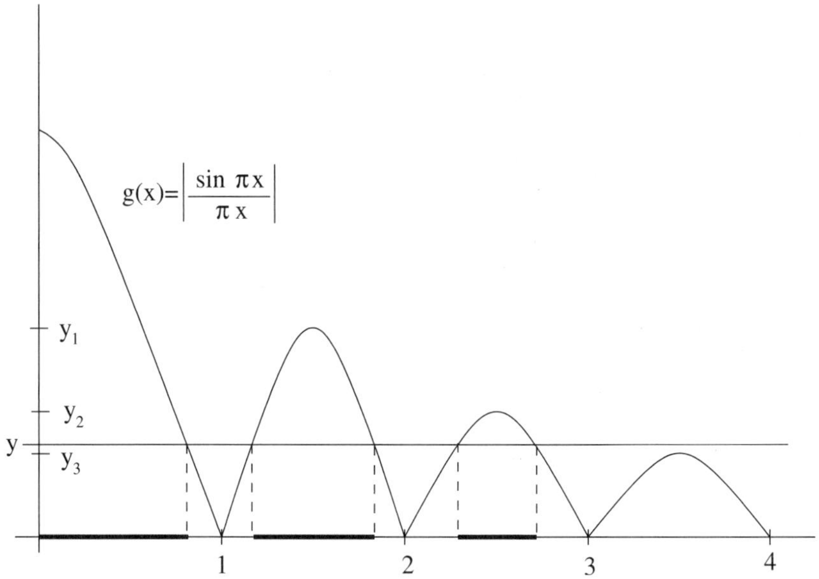

The function g.

The entire function $\sin(\pi x)/(\pi x)$ has roots at non-zero integers, so, by Euler's formula (see a more general Hadamard representation theorem for entire functions in [Lv, p. 26]),

$$g(x) = \prod_{k=1}^{\infty} \left|1 - \frac{x^2}{k^2}\right|.$$

For $x \in (0, 1)$, we can drop the absolute values and use the inequality $1 - x < \exp(-x)$:

$$g(x) = \prod_{k=1}^{\infty} (1 - x^2/k^2) \leq \prod_{k=1}^{\infty} \exp(-x^2/k^2)$$

$$= \exp(-\pi^2 x^2/6) < \exp(-\pi x^2/2) = f(x).$$

This means that $G(y) < F(y)$ on $(y_1, 1)$, because $g(x)$ can be greater than y_1 only on a subset of $(0,1)$. On the other hand, as shown in the proof of Lemma 7.3,

$$0 = \int_0^\infty f^2 - g^2 = \int_0^1 2y(F(y) - G(y))\, dy,$$

so F cannot be greater than G everywhere, and $F - G$ changes its sign somewhere on $(0,1)$. We need to prove that the change of sign occurs only once.

To do that, we prove that $F - G$ is an increasing function on $(0, y_1)$. Since both F and G are continuous decreasing functions, we need to prove that $|G'(y)| > |F'(y)|$ on each interval (y_{m+1}, y_m). Considering the level sets of the function g, one can see that

$$|G'(y)| = \sum \frac{1}{|g'(x)|},$$

where the sum is taken over all numbers x such that $g(x) = y$. There is one such number in the interval $(0,1)$ and two in each interval $(k, k+1)$, $k = 1, ..., m$.

For the roots $x \in (k, k+1)$ we have

$$|g'(x)| = \left| \frac{\cos(\pi x)}{x} - \frac{\sin(\pi x)}{\pi x^2} \right| \leq \frac{1}{x}\left(1 + \frac{|\sin(\pi(x-k))|}{\pi x}\right)$$

$$\leq \frac{1}{x}\left(1 + \frac{\pi(x-k)}{\pi k}\right) = \frac{1}{k}.$$

For the root $x \in (0,1)$,

$$|g'(x)| = \frac{\sin(\pi x) - \pi x \cos(\pi x)}{\pi x^2}$$

$$= \frac{1}{\pi x^2} \int_0^{\pi x} t \sin t\, dt \leq \frac{1}{\pi x^2} \int_0^{\pi x} t\, dt = \pi/2.$$

Thus, for every $y \in (y_{m+1}, y_m)$, we have

$$|G'(y)| \geq \frac{2}{\pi} + 2\sum_{k=1}^{m} k = \frac{2}{\pi} + m + m^2.$$

The function f is decreasing so

$$F(y) = f^{-1}(y) = \sqrt{\frac{2\ln(\frac{1}{y})}{\pi}}.$$

Therefore, for $y \in (y_{m+1}, y_m)$,

$$\frac{|G'(y)|}{|F'(y)|} = |G'(y)| y \sqrt{2\pi \ln(\frac{1}{y})} \geq \left(\frac{2}{\pi} + m + m^2\right) y \sqrt{2\pi \ln(\frac{1}{y})}.$$

The function $y\sqrt{2\pi \ln(1/y)}$ is increasing on $(0, y_1)$, since $y_1 < 1/\pi < 1/\sqrt{e}$. Also $y > y_{m+1} > 1/(\pi(m + 3/2))$, and we get

$$\frac{|G'(y)|}{|F'(y)|} > \frac{\frac{2}{\pi} + m + m^2}{\frac{3}{2} + m} \sqrt{\frac{2}{\pi} \ln \pi(m + \frac{3}{2})} > \sqrt{\frac{2}{\pi} \ln(\frac{5\pi}{2})}$$

($m \geq 1$ and $2/\pi + m^2 > 3/2$). The quantity in the right-hand side is greater than 1, which finishes the proof. ∎

Finding the minimal hyperplane section of the cube is easier. We follow here the approach of Hensley [He1].

LEMMA 7.5. *Let f be a non-negative non-increasing integrable continuous function on $[0,\infty)$ such that $\int_0^\infty t^2 f(t)dt < \infty$. Then*

$$3\int_0^\infty t^2 f(t)dt\,(f(0))^2 \geq \left(\int_0^\infty f(t)\,dt\right)^3. \tag{7.2}$$

PROOF. Suppose that $f(0) > 0$. Let $V = \int_0^\infty f$ and $a = V/f(0)$. Let g be a function on $[0,\infty)$ so that $g(t) = f(0)$ on $[0,a]$ and $g(t) = 0$ otherwise. Clearly, $\int_0^\infty g = \int_0^a g = V$. Since $g \geq f$ on the interval $[0,a]$, we have

$$\int_0^\infty t^2(f(t)-g(t))\,dt = \int_0^a t^2(f(t)-g(t))\,dt + \int_a^\infty t^2 f(t)\,dt$$

$$\geq a^2\int_0^a (f(t)-g(t))\,dt + a^2\int_a^\infty f(t)\,dt = a^2\int_0^\infty (f(t)-g(t))dt = 0.$$

Therefore,

$$\int_0^\infty t^2 f(t)\,dt \geq \int_0^\infty t^2 g(t)\,dt = f(0)\frac{1}{3}\frac{V^3}{(f(0))^3}.$$

□

THEOREM 7.6. *For every $\xi \in \mathbb{R}^n$,*

$$\mathrm{Vol}_{n-1}(Q_n \cap \xi^\perp) \geq 1$$

with equality for $\xi = (1,0,...,0)$.

PROOF. Fix $\xi \in S^{n-1}$. Let

$$f(t) = A_\xi(t) = \mathrm{Vol}_{n-1}(Q_n \cap \{\xi^\perp + t\xi\}).$$

By Brunn's theorem, Theorem 2.3, and by parity of f, the function f is non-increasing on $[0,\infty)$. We have $\int_0^\infty f = 1/2$ and

$$\int_0^\infty t^2 f(t)\,dt = \frac{1}{2}\int_{Q_n}(x,\xi)^2\,d\xi = \frac{1}{2}\int_{Q_n} x_1^2\,dx = \frac{1}{24}.$$

The result follows from Lemma 7.5, since $\mathrm{Vol}_{n-1}(Q_n \cap \xi^\perp) = f(0)$. □

7.2. The case $0 < q \leq 2$

In this section we consider the problem of finding the extremal central hyperplane sections of ℓ_q^n-balls with $0 < q \leq 2$. We prove that the minimal central hyperplane section of B_q^n is the one perpendicular to the vector $(1,1,..,1)$, and the maximal section is perpendicular to the vector $(1,0,...,0)$.

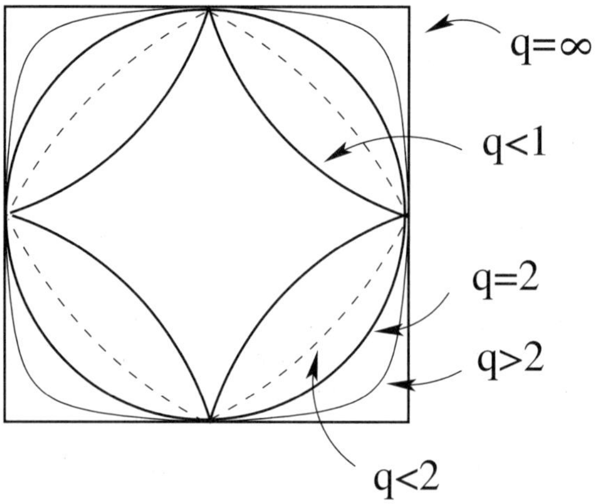

Unit balls of l_q^2.

THEOREM 7.7. *For $0 < q \le 2$, and any $\xi \in S^{n-1}$,*

$$\mathrm{Vol}(B_q^n \cap (\frac{1}{\sqrt{n}}, ..., \frac{1}{\sqrt{n}})^{\perp}) \le \mathrm{Vol}(B_q^n \cap \xi^{\perp}) \le \mathrm{Vol}(B_q^n \cap (1,0,...,0)^{\perp}).$$

PROOF. By Theorem 3.2, for every $\xi \in S^{n-1}$ and every $q > 0$,

$$\mathrm{Vol}_{n-1}(B_q^n \cap \xi^{\perp}) = \frac{q}{\pi(n-1)\Gamma((n-1)/q)} \int_0^{\infty} \prod_{k=1}^{n} \gamma_q(t\xi_k)\, dt,$$

where γ_q is the Fourier transform of the function $\exp(-|\cdot|^q)$ on $\mathbb{R}$.

As noted in [MeP], to prove the theorem, it is enough to show that the function $\ln(\gamma_q(\sqrt{\cdot}))$ is convex on $[0,\infty)$. In fact, convexity of this function implies that, for any $0 < \xi_1 < \eta_1 < \eta_2 < \xi_2$ with $\xi_1^2 + \xi_2^2 = \eta_1^2 + \eta_2^2 = 1$, one has

$$\gamma_q(t\xi_1)\gamma_q(t\xi_2) \ge \gamma_q(t\eta_1)\gamma_q(t\eta_2)$$

for all $t > 0$. This means, in particular, that the integral above is minimal when all the coordinates of the vector ξ are equal, and this integral is maximal when one of the coordinates of ξ equals 1 and the rest are equal to zero.

To prove that the function $\ln(\gamma_q(\sqrt{\cdot}))$ is convex on $[0,\infty)$, we use the representation (2.51). There exists a finite measure $\mu_{q/2}$ on $[0,\infty)$ so that, for every $\xi \in \mathbb{R}$,

$$\gamma_q(\xi) = \sqrt{\pi} \int_0^{\infty} t^{-1/2} \exp(-\frac{\xi^2}{4t})\, d\mu_{q/2}(t).$$

By the Cauchy-Schwartz inequality, for any $\xi_1, \xi_2 > 0$,

$$\gamma_q^2\left(\sqrt{\frac{\xi_1+\xi_2}{2}}\right) = \pi\left(\int_0^{\infty} t^{-1/2} \exp(-\frac{\xi_1}{8t}) \exp(-\frac{\xi_2}{8t})\, d\mu_{q/2}(t)\right)^2$$

$$\le \pi \int_0^{\infty} t^{-1/2} \exp(-\frac{\xi_1}{4t})\, d\mu_{q/2}(t) \int_0^{\infty} t^{-1/2} \exp(-\frac{\xi_2}{4t})\, d\mu_{q/2}(t)$$

$$= \gamma_q(\sqrt{\xi_1})\gamma_q(\sqrt{\xi_2}),$$

which proves our claim. □

7.3. Remarks and further results

It is quite surprising that the maximal section of the cube was found only in the 1980's. In 1979, Hensley [He1] proved that hyperplane sections of the cube are bounded from above by a constant not depending on the dimension. Theorem 7.1 giving the exact value $\sqrt{2}$ for the maximal volume of hyperplane sections of the cube was proved by Ball [Ba1] in 1986. Different proofs of Ball's integral inequality, Lemma 7.4, were later found by Nazarov and Podkorytov [NP] and by Oleszkiewicz and Pelczynski [OP]. The proof here follows that from [NP].

The minimal hyperplane section of the cube was first found by Hadwiger [Ha1]. Other proofs were given later, for example Vaaler [Va] proved that the minimal volume of a section of any codimension is 1. The proof that we presented here is due to Hensley [He1].

Ball [Ba3] used the inequality of Lemma 7.4 and the Brascamp-Lieb inequality in place of Hölder's inequality to prove that the maximal volume of a section of the cube by a subspace of codimension $k < n/2$ is $2^{k/2}$. It is also proved in [Ba3] that if $n-k$ divides n, then the maximal volume of $(n-k)$-dimensional sections of the cube is $(\frac{n}{n-k})^{(n-k)/2}$.

The question of finding extremal slabs in the cube was asked by V. Milman and considered in the paper [BK]. It was proved that, for $t < 3/4$, the slab parallel to a face has the minimal volume among all symmetric slabs with width t. Note that the constant $3/4$ is not the best possible even in dimension 2. The proof uses a technique generalizing that from Hensley's proof of the minimal hyperplane section, Lemma 7.5 and Theorem 7.6. For large width, the slab orthogonal to the main diagonal appears to be asymptotically minimal. The proof is based on certain concavity properties of the Laplace transform and on several limit theorems from probability: the central limit theorem and classical principles of moderate and large deviations.

Oleszkiewicz and Pelczynski [OP] proved a complex version of Ball's result. In place of Ball's inequality, they have an inequality involving Bessel functions.

Meyer and Pajor [MeP] proved in 1988 that, for every $1 \le q \le p \le \infty$ and every k-dimensional subspace H of $\mathbb{R}^n$,

$$\frac{\mathrm{Vol}(B_p^n \cap H)}{\mathrm{Vol}(B_p^k)} \ge \frac{\mathrm{Vol}(B_q^n \cap H)}{\mathrm{Vol}(B_q^k)}.$$

Putting $q = 2$, we see that for any $p \ge 2$ the minimal k-dimensional section of B_p^n is B_p^k. On the other hand, putting $p = 2$, we see that for $1 \le q \le 2$, the maximal k-dimensional section of B_q^n is B_q^k (this was later proved by a different method by Ball [Ba8]). In the case $0 < q < 1$ the latter result was proved by Caetano [Cae] and Barthe [Bar]. This left open the case of minimal sections of B_q^n with $0 < q < 2$ and the case of maximal sections of B_q^n with $q > 2$.

Meyer and Pajor also proved in [MeP] that the minimal central hyperplane section of B_1^n is the one perpendicular to the main diagonal, and they conjectured that the result is the same for any B_q^n with $0 < q \le 2$. This conjecture was proved in [K7], and the result is presented in Theorem 7.7 in Section 7.2. The proof gives the minimal and maximal sections simultaneously. A complex version of this result was given in [KZ].

The problem of finding the maximal hyperplane section of B_q^n with $q > 2$ is still open and seems to be very difficult. Oleszkiewicz [O] has recently shown that, unlike the case of the cube, the answer must depend on both q and n.

CHAPTER 8

Projections and the Fourier Transform

In this chapter we describe the Fourier analytic approach to projections of convex bodies. The results are similar to those for sections. We start with the main formula, expressing the volume of hyperplane projections in terms of the Fourier transform of the curvature function.

8.1. A formula for the volume of hyperplane projections

Let μ be a finite Borel measure on the unit sphere S^{n-1}. We extend μ to a homogeneous distribution of degree $-n-1$. A distribution μ_e is called the extended measure of μ if, for every even test function $\phi \in S(\mathbb{R}^n)$,

$$\langle \mu_e, \phi \rangle = \frac{1}{2} \int_{S^{n-1}} \langle r^{-2}, \phi(r\xi) \rangle d\mu(\xi). \tag{8.1}$$

In most cases we are only interested in test functions supported outside of the origin, for which $\langle r^{-2}, \phi(r\xi) \rangle = \int_{\mathbb{R}} r^{-2} \phi(r\xi)\, dr$.

If μ is absolutely continuous with the density $g \in L_1(S^{n-1})$, we define the extension $g(x)$, $x \in \mathbb{R}^n \setminus \{0\}$, as a homogeneous function of degree $-n-1$, $g(x) = |x|_2^{-n-1} g(x/|x|_2)$, and identify μ_e with g.

The following fact is well known; see for example [Se, formula 4.7], [K6, Lemma 2] or [KRZ, Theorem 1]. We have already proved a more general fact in Lemma 6.9 (note that here we are computing the Fourier transform of the norm of a subspace of L_1) and a similar result in Corollary 3.15, so we omit the proof here. The exact value of the constant can be computed in Lemma 6.9 using the constant from Lemma 2.23.

THEOREM 8.1. *The Fourier transform $\widehat{\mu_e}$ is a continuous homogeneous of degree 1 even function on $\mathbb{R}^n$. For every $\theta \in S^{n-1}$,*

$$\widehat{\mu_e}(\theta) = -\frac{\pi}{2} \int_{S^{n-1}} |(\theta, y)| d\mu(y).$$

The following formula is our main tool in this chapter. It immediately follows from the Cauchy formula (2.14) and Theorem 8.1. We denote by $S_e(L)$ the extended measure of the surface area measure $S(L, \cdot)$. Also we extend the curvature function f_L to $\mathbb{R}^n \setminus \{0\}$ as a homogeneous function of degree $-n-1$.

THEOREM 8.2. *Let L be a convex origin-symmetric body in $\mathbb{R}^n$. Then*

$$\widehat{S_e(L)}(\theta) = -\pi \mathrm{Vol}_{n-1}\left(L\middle|\theta^\perp\right), \qquad \forall \theta \in S^{n-1}.$$

In particular, if the body L has a curvature function, then

$$\widehat{f_L}(\theta) = -\pi \mathrm{Vol}_{n-1}\left(L\middle|\theta^\perp\right), \qquad \forall \theta \in S^{n-1}. \tag{8.2}$$

8.2. Extremal hyperplane projections of ℓ_q-balls

The case of hyperplane projections of the cube is quite simple. Using the Cauchy formula (all the facets of the unit cube have the same volume, and their normal vectors are parallel to the axes), we get

$$\mathrm{Vol}_{n-1}\left(B_\infty^n\Big|\theta^\perp\right) = \mathrm{Vol}_{n-1}(B_\infty^{n-1})\left(\sum_{i=1}^n |\theta_i|\right).$$

Then

$$\mathrm{Vol}\left(B_\infty^n\Big|(1,0,\dots,0)^\perp\right) \le \mathrm{Vol}\left(B_\infty^n\Big|\theta^\perp\right)$$
$$\le \mathrm{Vol}\left(B_\infty^n\Big|(\frac{1}{\sqrt{n}},\dots,\frac{1}{\sqrt{n}})^\perp\right).$$

The case of projections of B_1^n is more difficult. Consider vectors $\delta = (\delta_1,\dots,\delta_n)$, where δ_i can be chosen as $+1$ or -1. Then $\frac{1}{\sqrt{n}}\delta$ are normal vectors to the facets of B_1^n. Let C_n be the volume of a facet of B_1^n. Then again the Cauchy formula implies

$$\mathrm{Vol}_{n-1}\left(B_1^n\Big|\theta^\perp\right) = \frac{1}{\sqrt{n}}C_n\left(\sum_\delta |(\theta,\delta)|\right).$$

Finding the maximal and minimal values of $\sum_\varepsilon |(\theta,\delta)|$ is related to sharp constants in the Khinchine inequality for independent symmetric Bernoulli random variables. These constants were found by Szarek [Sz]:

$$\mathrm{Vol}\left(B_1^n\Big|(\frac{1}{\sqrt{2}},\frac{1}{\sqrt{2}},0,\dots,0)^\perp\right) \le \mathrm{Vol}\left(B_1^n\Big|\theta^\perp\right) \le \mathrm{Vol}\left(B_1^n\Big|(1,0,\dots,0)^\perp\right).$$

Comparing the results on extremal sections and projections of B_1^n and its polar B_∞^n, we see that the answers for projections are "dual" to the answers for sections.

The result of Barthe and Naor [BN] on the extremal hyperplane projections of the balls B_p^n with $p \ge 2$ is as follows:

THEOREM 8.3. *Let $p \ge 2$ and $\theta \in S^{n-1}$. Then*

$$\mathrm{Vol}_{n-1}\left(B_p^n\Big|(1,0,...,0)^\perp\right) \le \mathrm{Vol}_{n-1}\left(B_p^n\Big|\theta^\perp\right) \le \mathrm{Vol}_{n-1}\left(B_p^n\Big|(\frac{1}{\sqrt{n}},...,\frac{1}{\sqrt{n}})^\perp\right).$$

The proof in [BN] is based on a formula similar to (3.2): for every $\xi \in S^{n-1}$,

$$\mathrm{Vol}_{n-1}\left(B_p^n\Big|\xi^\perp\right) = \frac{2^n\Gamma^n(1/p)}{\pi p^{n-1}\Gamma(n-\frac{n-1}{p^*})}\int_0^\infty \frac{1-\prod_{k=1}^n \frac{p^*}{2\Gamma(1/p)}\beta_{p^*}(t\xi_k)}{t^2}\,dt, \tag{8.3}$$

where $1/p + 1/p^* = 1$ and β_{p^*} is the Fourier transform of the function $|\cdot|^{p^*-2}e^{-|\cdot|^{p^*}}$ on $\mathbb{R}$. The proof of this formula in [BN] uses probabilistic arguments. The rest of the proof of Theorem 8.3 is similar to that for sections. For $p \ge 2$, the function $\beta_{p^*}(\sqrt{\cdot})$ is log-convex on $[0,\infty)$, which is proved in a way similar to the case of the function γ_q in Theorem 7.7. Theorem 8.3 immediately follows from this and (8.3).

In this section we show that formula (8.3) is a particular case of the formula (8.2) and holds for every $p > 1$. This allows us to completely unify methods in the section and projection cases. We do that by computing the Fourier transform of the curvature function of l_p-balls. First, let us get an expression for the curvature function using methods of differential geometry from [Th].

LEMMA 8.4. *The curvature function of the ball* B_p^n *is*

$$f_{B_p^n}(\theta) = (p^* - 1)^{n-1} \left(\prod_{i=1}^n |\theta_i|^{p^*-2} \right) \|\theta\|_{p^*}^{(n-1)-np^*}, \qquad \theta \in S^{n-1}.$$

PROOF. To find $f_{B_p^n}(\theta)$, we first compute the Gauss curvature K_p of B_p^n. Let $F(x_1, \dots, x_n) = \sum_{i=1}^n |x_i|^p$. Then the boundary $\partial B_p^n = \{x \in \mathbb{R}^n : F(x_1, \dots, x_n) = 1\}$. Let

$$Z(x) = -\frac{1}{p} \nabla F(x) = - \left(|x_1|^{p-1} \mathrm{sgn}(x_1), \dots, |x_n|^{p-1} \mathrm{sgn}(x_n) \right).$$

We use the following formula for the Gauss curvature (see, for example, [Th, p. 89]:

$$K(x) = (-1)^{n-1} \det \begin{pmatrix} \nabla_{v_1} Z \\ \vdots \\ \nabla_{v_{n-1}} Z \\ Z(x) \end{pmatrix} \Big/ |Z(x)|^{n-1} \det \begin{pmatrix} v_1 \\ \vdots \\ v_{n-1} \\ Z(x) \end{pmatrix}. \tag{8.4}$$

Here vectors $v_1, \dots, v_{n-1}$ can be chosen as any basis in the tangent space at the corresponding point, and

$$\nabla_{v_i} Z = (\nabla_{v_i} Z_1, \dots, \nabla_{v_i} Z_n) = ((v_i, \nabla Z_1), \dots, (v_i, \nabla Z_n)).$$

We choose $v_1, \dots, v_{n-1}$ in the following way: for every k, $1 \le k \le n-1$,

$$v_k = (|x_{k+1}|^{p-1} \mathrm{sgn}(x_{k+1}), \underbrace{0, \dots, 0}_{k-1 \text{ times}}, -|x_1|^{p-1} \mathrm{sgn}(x_1), 0, \dots, 0,).$$

Then

$$\nabla_{v_k} Z(x) = (p-1) |x_1 \, x_{k+1}|^{p-2} (-x_{k+1}, 0, \dots, 0, \underbrace{x_1}_{(k+1)s \text{ place}}, 0, \dots, 0,).$$

Substituting this in (8.4) (computation of the determinants is standard), we get

$$K_p(x) = (p-1)^{n-1} |Z(x)|_2^{-n-1} \prod_{i=1}^n |x_i|^{p-2} \tag{8.5}$$

for almost all $x \in \partial B_p^n$.

Let $\theta = Z(x)/|Z(x)|_2$. Recall that f_K is the reciprocal Gauss curvature, viewed as a function of the unit normal vector (see [Sch3, p. 419], so

$$f_{B_p^n}(\theta) = 1/K_p(Z(x)/|Z(x)|_2).$$

Since $\theta = Z(x)/|Z(x)|_2$, we see that θ and $Z(x)$ are collinear vectors in $\mathbb{R}^n$, i.e. $\theta = \lambda Z(x)$ for some $\lambda > 0$. Next

$$x_i = \lambda^{1/(p-1)} |\theta_i|^{1/(p-1)} \mathrm{sgn}(\theta_i),$$

and using $\|x\|_p = 1$, we get

$$\lambda = \frac{1}{\left(\sum_{i=1}^n |\theta_i|^{p/(p-1)} \right)^{(p-1)/p}}.$$

Finally

$$x_i = -\left(|\theta_i|^{1/(p-1)}\operatorname{sgn}(\theta_i)\right) / \left(\sum_{i=1}^{n} |\theta_i|^{p/(p-1)}\right)^{1/p}.$$

To finish the proof, we substitute the last expression into (8.5) and use $p-1 = (p^*-1)^{-1}$. □

We now prove formula (8.3) by computing the Fourier transform of the function appearing in the previous lemma.

THEOREM 8.5. *Formula (8.3) holds for every $p > 1$.*

PROOF. According to Theorem 8.2, we have to compute the Fourier transform of $f_{B_p^n}(x) = |x|_2^{-n-1} f_{B_p^n}(x/|x|_2)$, $x \in \mathbb{R}^n \setminus \{0\}$. From the definition of the Gamma function

$$\|x\|_{p^*}^{-q} = \frac{p^*}{\Gamma(q/p^*)} \int_0^\infty y^{q-1} e^{-y^{p^*}(|x_1|^{p^*}+\cdots+|x_1|^{p^*})} dy \tag{8.6}$$

for any $q > 0$ and $x \in \mathbb{R}^n \setminus \{0\}$.

Let $\phi \in \mathcal{S}(\mathbb{R}^n)$ be even and such that $0 \notin \operatorname{supp}(\widehat{\phi})$. Using Lemma 8.4 and (8.6), we get

$$\langle \widehat{f_{B_p^n}}, \phi\rangle = \langle f_{B_p^n}, \widehat{\phi}\rangle$$

$$= \frac{(p^*-1)^{n-1}p^*}{\Gamma(q/p^*)} \int_0^\infty y^{q-1} \left(\int_{\mathbb{R}^n} \prod_{k=1}^n |x_k|^{p^*-2} e^{-y^{p^*}|x_k|^{p^*}} \widehat{\phi}(x)dx \right) dy$$

$$= \frac{(p^*-1)^{n-1}p^*}{\Gamma(q/p^*)} \int_0^\infty y^{q-1} \left(\int_{\mathbb{R}^n} \left(\prod_{k=1}^n |x_k|^{p^*-2} e^{-y^{p^*}|x_k|^{p^*}} \right)^\wedge (\xi)\phi(\xi)d\xi \right) dy$$

$$= \frac{(p^*-1)^{n-1}p^*}{\Gamma(q/p^*)} \int_0^\infty y^{q-1} \left(\int_{\mathbb{R}^n} \prod_{k=1}^n \left(|t|^{p^*-2} e^{-y^{p^*}|t|^{p^*}} \right)^\wedge (\xi_k)\phi(\xi)d\xi \right) dy.$$

This gives

$$\langle \widehat{f_{B_p^n}}, \phi\rangle = \frac{(p^*-1)^{n-1}p^*}{\Gamma(q/p^*)} \int_0^\infty y^{q-1+(1-p^*)n} \left(\int_{\mathbb{R}^n} \prod_{k=1}^n \beta_{p^*}(\frac{\xi_k}{y})\, \phi(\xi)d\xi \right) dy.$$

Now replacing q by $np^* - (n-1)$ and using the fact that $0 \notin \operatorname{supp}(\widehat{\phi})$ implies $\int_{\mathbb{R}^n} \phi(\xi)d\xi = 0$, we obtain

$$\langle \widehat{f_{B_p^n}}, \phi\rangle = \frac{(p^*-1)^{n-1}p^*}{\Gamma(n-(n-1)/p^*)} \int_0^\infty \left(\int_{\mathbb{R}^n} \left(\prod_{k=1}^n \beta_{p^*}(\frac{\xi_k}{y}) - \beta_{p^*}^n(0) \right) \phi(\xi)d\xi \right) dy$$

$$= \frac{(p^*-1)^{n-1}p^*}{\Gamma(n-(n-1)/p^*)} \int_{\mathbb{R}^n} \phi(\xi) \left(\int_0^\infty \prod_{k=1}^n \beta_{p^*}(\frac{\xi_k}{y}) - \beta_{p^*}^n(0)dy \right) d\xi,$$

where

$$\beta_{p^*}(0) = \int_{-\infty}^{\infty} |u|^{p^*-2} e^{-|u|^{p^*}} du = \frac{2}{p^*} \int_0^{\infty} v^{-1/p^*} e^{-v} dv = \frac{2}{p^*} \Gamma(1 - 1/p^*).$$

Making the substitution $\xi_k/y = t$, we finally get

$$\langle \widehat{f_{B_p^n}}, \phi \rangle = \frac{(p^*-1)^{n-1} p^*}{\Gamma(n-(n-1)/p^*)} \int_{\mathbb{R}^n} \phi(\xi) \left(\int_0^{\infty} \frac{\prod_{k=1}^n \beta_{p^*}(t\xi_k) - \beta_{p^*}^n(0)}{t^2} dt \right) d\xi$$

$$= \frac{2^n \Gamma^n(1/p)}{p^{n-1} \Gamma(n-(n-1)/p^*)} \int_{\mathbb{R}^n} \phi(\xi) \left(\int_0^{\infty} \frac{\prod_{k=1}^n \frac{p^*}{2\Gamma(1/p)} \beta_{p^*}(t\xi_k) - 1}{t^2} dt \right) d\xi.$$

Since ϕ is an arbitrary even test function with $0 \notin \operatorname{supp}(\hat{\phi})$, the distributions in the left- and right-hand sides of the latter equality are equal (they are both even homogeneous functions of degree 1). Now the result follows from Theorem 8.2. □

8.3. Projection bodies

Let K be an origin-symmetric convex body in $\mathbb{R}^n$. The *projection body* ΠK of K is defined as an origin-symmetric convex body in $\mathbb{R}^n$ whose support function in every direction is equal to the volume of the hyperplane projection of K to this direction: for every $\theta \in S^{n-1}$,

$$h_{\Pi K}(\theta) = \mathrm{Vol}_{n-1}(K|\theta^{\perp}) = \frac{1}{2} \int_{S^{n-1}} |(\theta, v)| dS(K, v), \tag{8.7}$$

where the second equality is the Cauchy formula (2.14).

We extend $h_{\Pi K}$ to a homogeneous function of degree 1 on $\mathbb{R}^n$. It immediately follows from Theorem 8.1 that the Fourier transform of $h_{\Pi K}$ is equal (up to a constant) to the extended surface area measure of K :

$$\widehat{h_{\Pi K}} = -(2\pi)^{n-1} S_e(K). \tag{8.8}$$

Conversely, if L is an origin-symmetric convex body so that $-(2\pi)^{-n+1}\widehat{h_L} = \mu_e$ is an extended measure of some even finite Borel measure μ on S^{n-1}, then, again by Theorem 8.1, one has

$$h_L(\theta) = \frac{1}{2} \int_{S^{n-1}} |(\theta, v)| d\mu(v).$$

Since L is an n-dimensional body, the measure μ cannot be supported in any great subsphere of S^{n-1}, and, by Minkowski's existence theorem, Theorem 2.9, the measure μ is the surface area measure of some origin-symmetric convex body K. By Cauchy's formula (2.14), we have $L = \Pi K$. We have proved the following Fourier analytic characterization of projection bodies (the last statement follows from Theorem 6.10 and the fact that h_L is the norm of the dual body):

THEOREM 8.6. *An origin-symmetric convex body L in $\mathbb{R}^n$ is the projection body of some origin-symmetric convex body if and only if there exists a measure μ on S^{n-1} so that*

$$\widehat{h_L} = -(2\pi)^{n-1}\mu_e. \tag{8.9}$$

The measure μ serves as the surface area measure of an origin-symmetric convex body K so that $L = \Pi K$.

Also, L is a projection body if and only if the polar body L^* is the unit ball of a subspace of L_1.

Suppose now that an origin-symmetric convex body L in $\mathbb{R}^n$ has the property that h_L is an infinitely differentiable function on the sphere S^{n-1}. This means that the polar body L^* is infinitely smooth (recall that $h_L = \|\cdot\|_{L^*}$). Putting $k = n$, $D = L^*$ in the formulas of Theorem 3.18, we get that if n is even, then, for every $\xi \in S^{n-1}$,

$$\widehat{h_L}(\xi) = (-1)^{1+n/2}\pi A^{(n)}_{L^*,\xi}(0) \tag{8.10}$$

and if n is odd, then, for every $\xi \in S^{n-1}$,
(8.11)

$$\widehat{h_L}(\xi) = (-1)^{(n-1)/2}2n!\int_0^\infty \frac{A_{L^*,\xi}(z) - A_{L^*,\xi}(0) - \dots - A^{(n-1)}_{L^*,\xi}(0)\frac{z^{n-1}}{(n-1)!}}{z^{n+1}}\,dz.$$

We can now characterize projection bodies in terms of sections of the polar body. Compare this result with a similar characterization of intersection bodies in Corollary 4.3.

THEOREM 8.7. *Let L be an origin-symmetric convex body in $\mathbb{R}^n$ so that h_L is infinitely differentiable on S^{n-1}. The body L is a projection body of some convex body K if and only if for every $\xi \in S^{n-1}$,*
(i) if n is even

$$(-1)^{n/2}A^{(n)}_{L^*,\xi}(0) \geq 0;$$

(ii) if n is odd

$$(-1)^{(n+1)/2}\int_0^\infty \frac{A_{L^*,\xi}(z) - A_{L^*,\xi}(0) - \dots - A^{(n-1)}_{L^*,\xi}(0)\frac{z^{n-1}}{(n-1)!}}{z^{n+1}}\,dz \geq 0.$$

We end this section by proving a version of Parseval's formula on the sphere that will be used in the solution of Shephard's problem. The proof that we present here is based on spherical harmonics and is very similar to the proof of Lemma 5.10. This proof is quite short, but somewhat artificial, in the sense that it is not directly clear that the spherical version of Parseval's formula is the result of projecting Parseval's formula on $\mathbb{R}^n$ to the sphere. A more natural (but longer) proof of this lemma, essentially repeating the proof of Lemma 3.22, can be found in [KRZ].

LEMMA 8.8. *Let K, L be origin-symmetric convex bodies in $\mathbb{R}^n$ so that the support function of K is infinitely smooth on the sphere and the curvature function of L exists and is a continuous function on the sphere. Then*

$$\int_{S^{n-1}} \widehat{h_K}(\xi)\widehat{f_L}(\xi)\,d\xi = (2\pi)^n\int_{S^{n-1}} h_K(x)f_L(x)\,dx.$$

PROOF. Assume first that n is an even integer. By (8.10) and Lemma 5.9 (extended analytically to $q = n$), for every spherical harmonic $Y_m \in P_m$ where m is an even non-negative integer,

$$(8.12) \qquad \left(\widehat{h_K}, Y_m\right) = \left((-1)^{1+n/2}\pi A^{(n)}_{K^*,\theta}(0), Y_m(\theta)\right)$$

$$= \frac{(-1)^{1+n/2}\pi 2^{n+1}\pi^{n/2-1}\sin(\pi(m+n+1)/2)\Gamma((m+n+1)/2)}{(-1)\Gamma((m-1)/2)}$$

$$\times \int_{S^{n-1}} \|\theta\|_{K^*} Y_m(\theta)d\theta.$$

On the other hand, if we use (8.2) and then apply the Funk-Hecke theorem (5.7) and (5.9) with $q = 1$ to the Cauchy formula (2.14), we get

$$(8.13) \qquad \left(\widehat{f_L}, Y_m\right) = -\pi\left(\mathrm{Vol}_{n-1}(L\big|\theta^\perp), Y_m(\theta)\right)$$

$$= -\frac{\pi}{2}\left(\int_{S^{n-1}} |(\theta,\xi)| f_L(\xi)d\xi, Y_m(\theta)\right)$$

$$= -\frac{\pi}{2}\int_{S^{n-1}} f_L(\xi)\ \left(|(\theta,\xi)|, Y_m(\theta)\right)\ d\xi$$

$$= -\frac{\pi}{2}\frac{\pi^{n/2-1}\Gamma(2)\sin(\pi(m-1)/2)\Gamma((m-1)/2)}{\Gamma((m+n+1)/2)}\int_{S^{n-1}} f_L(\xi)Y_m(\xi)d\xi.$$

Multiplying (8.12) and (8.13), we get

$$\left(\widehat{h_K}, Y_m\right)\left(\widehat{f_L}, Y_m\right) = (2\pi)^n \int_{S^{n-1}} \|\theta\|_{K^*} Y_m(\theta)d\theta \int_{S^{n-1}} f_L(\theta)Y_m(\theta)d\theta.$$

Applying the latter equality to each function Y_{mj} from an orthonormal basis of spherical harmonics in $L_2(S^{n-1})$ (note that for odd m the coefficients are equal to zero because of the symmetry), we get the result (note also that $\|\theta\|_{K^*} = h_K(\theta)$).

If n is an odd integer, the proof is essentially the same. Instead of using Lemma 5.9 directly, one has to use a special version of it for odd numbers q mentioned after Theorem 5.12, and then use (8.11) instead of (8.10). □

8.4. The Shephard problem

The Shephard problem (see [Sh]) reads as follows. Let K, L be convex origin-symmetric bodies in $\mathbb{R}^n$ and suppose that, for every $\theta \in S^{n-1}$,

$$(8.14) \qquad \mathrm{Vol}_{n-1}\left(K\big|\theta^\perp\right) \le \mathrm{Vol}_{n-1}\left(L\big|\theta^\perp\right).$$

Does it follow that

$$(8.15) \qquad \mathrm{Vol}_n(K) \le \mathrm{Vol}_n(L)?$$

Clearly, the answer is affirmative in dimension 2, because in this case

$$\mathrm{Vol}_{n-1}\left(K\big|\theta^\perp\right) = 2h_K(\theta) = 2\|\theta\|_{K^*},$$

so condition (8.14) means that $L^* \subset K^*$ and $K \subset L$ (note that symmetry is essential here, because the answer to Shephard's question in dimension 2 is negative in the non-symmetric case).

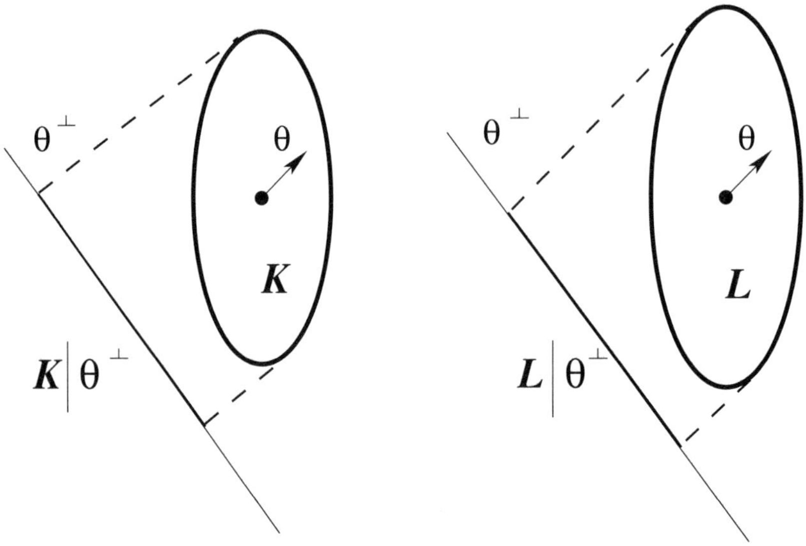

In dimensions 3 and higher the volume of hyperplane projections no longer equals the support function.

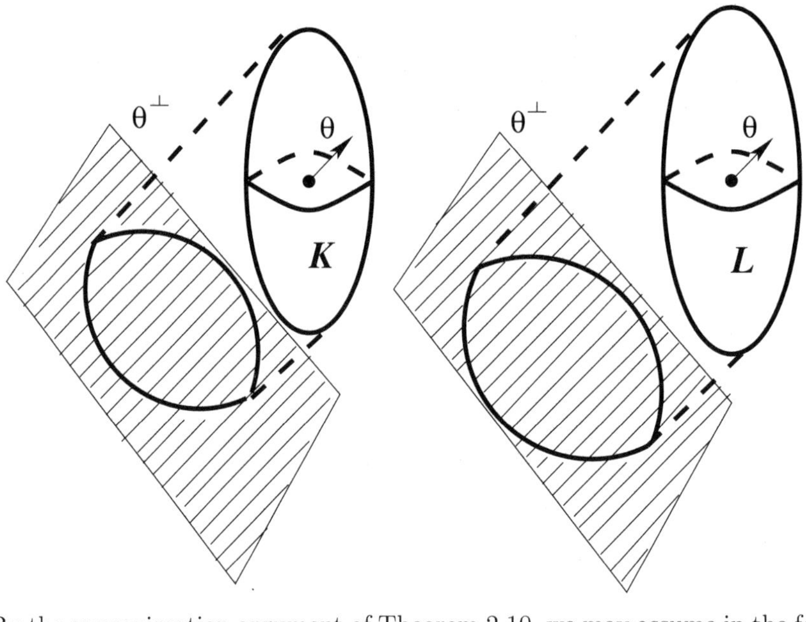

By the approximation argument of Theorem 2.10, we may assume in the formulation of Shephard's problem that the bodies K and L are such that their support functions h_K, h_L are infinitely smooth functions on $\mathbb{R}^n \setminus \{0\}$. Using Lemma 3.16, we get in this case that the Fourier transforms $\widehat{h_K}$, $\widehat{h_L}$ are the extensions of infinitely differentiable functions on the sphere to homogeneous distributions on $\mathbb{R}^n$ of degree $-n-1$. Moreover, by a similar approximation argument (see also [GrZ, Section 5]) we may assume that our bodies have absolutely continuous surface area measures. Therefore, in the rest of this section, K and L are convex symmetric bodies with infinitely smooth support functions and absolutely continuous surface area measures.

The following theorem is a projection analog of Lutwak's connections between intersection bodies and the Busemann-Petty problem.

THEOREM 8.9.
(i) If a body L is such that $\widehat{h_L} \leq 0$ for all $\theta \in S^{n-1}$, then the Shephard problem has an affirmative answer for this L and any K.
(ii) If $f_K(\theta) > 0$ for every $\theta \in S^{n-1}$ and $\widehat{h_K}$ is positive on an open subset of S^{n-1}, then there exists a body L that gives together with K a counterexample in Shephard's problem.

This theorem will follow from the next two lemmas and the fact that the condition

$$\mathrm{Vol}_{n-1}\left(K\middle|\theta^{\perp}\right) \leq \mathrm{Vol}_{n-1}\left(L\middle|\theta^{\perp}\right)$$

is equivalent (see Theorem 8.2) to

$$\widehat{f_K}(\theta) \geq \widehat{f_L}(\theta), \qquad \theta \in S^{n-1}.$$

LEMMA 8.10. *If $\widehat{h_L}(\theta) \leq 0$ and $\widehat{f_K}(\theta) \geq \widehat{f_L}(\theta)$, $\forall \theta \in S^{n-1}$, then*

$$\mathrm{Vol}_n(K) \leq \mathrm{Vol}_n(L).$$

PROOF. From $\widehat{f_K}(\theta) \geq \widehat{f_L}(\theta)$ and $\widehat{h_L}(\theta) \leq 0$ we get

$$\int_{S^{n-1}} \widehat{h_L}(\theta)\widehat{f_K}(\theta)d\theta \leq \int_{S^{n-1}} \widehat{h_L}(\theta)\widehat{f_L}(\theta)d\theta = (*).$$

Using Parseval's formula on the sphere, Lemma 8.8, and formula (2.16), we get that

$$(*) = (2\pi)^n \int_{S^{n-1}} h_L(\theta)f_L(\theta)d\theta = n(2\pi)^n \mathrm{Vol}_n(L).$$

On the other hand,

$$\int_{S^{n-1}} \widehat{h_L}(\theta)\widehat{f_K}(\theta)d\theta = (2\pi)^n \int_{S^{n-1}} h_L(\theta)f_K(\theta)d\theta = n(2\pi)^n V_1(K, L),$$

where $V_1(K, L)$ is the mixed volume. Thus,

$$V_1(K, L) \leq \mathrm{Vol}_n(L).$$

Now we apply the first Minkowski inequality (2.8):

$$V_1(K, L) \geq \mathrm{Vol}_n(L)^{\frac{1}{n}} \mathrm{Vol}_n(K)^{\frac{n-1}{n}}$$

to get

$$\mathrm{Vol}_n(L) \geq \mathrm{Vol}_n(L)^{\frac{1}{n}} \mathrm{Vol}_n(K)^{\frac{n-1}{n}},$$

or

$$\mathrm{Vol}_n(L) \geq \mathrm{Vol}_n(K).$$

□

LEMMA 8.11. *Let K be such that $f_K(\theta) > 0$, $\forall \theta \in S^{n-1}$. If $\widehat{h_K}$ is positive on an open subset of S^{n-1}, then there exists a convex symmetric body L in $\mathbb{R}^n$, such that*

$$\widehat{f_K} \geq \widehat{f_L},$$

but

$$\mathrm{Vol}_n(K) > \mathrm{Vol}_n(L).$$

PROOF. Let $\Omega = \{\theta \in S^{n-1} : \widehat{h_K}(\theta) > 0\}$ and let $v \in C^\infty(S^{n-1})$ be a non-positive even function supported on Ω, where v is not identically zero. We extend v to a homogeneous function $v(\theta)r$ of degree 1 on $\mathbb{R}^n$. Then the Fourier transform of $v(\theta)r$ is a homogeneous function of degree $-n-1$: $\widehat{v(\theta)r} = g(\theta)r^{-n-1}$, where g is an infinitely smooth function on S^{n-1} (see Lemma 3.16).

Since g is bounded on S^{n-1}, one can choose a small $\varepsilon > 0$ so that, for every $\theta \in S^{n-1}$ and $r > 0$,

$$f_L(r\theta) = f_K(r\theta) + \varepsilon g(\theta)r^{-n-1} > 0.$$

By the Minkowski existence theorem, Theorem 2.9, f_L defines a convex symmetric body $L \in \mathbb{R}^n$. By the definition of the function v,

$$\widehat{f_L}(r\theta) = \widehat{f_K}(r\theta) + \varepsilon v(\theta)r \le \widehat{f_K}(r\theta).$$

Next, since v is supported and is non-positive in the set where $\widehat{h_K} > 0$,

$$\int_{S^{n-1}} \widehat{h_K}(\theta)\widehat{f_L}(\theta)d\theta = \int_{S^{n-1}} \widehat{h_K}(\theta)\widehat{f_K}(\theta)d\theta + \int_{S^{n-1}} \widehat{h_K}(\theta)\varepsilon v(\theta)d\theta$$

$$< \int_{S^{n-1}} \widehat{h_K}(\theta)\widehat{f_K}(\theta)d\theta = (2\pi)^n \int_{S^{n-1}} h_K(\theta)f_K(\theta)d\theta = n(2\pi)^n \mathrm{Vol}_n(K).$$

But again by Parseval's formula, Lemma 8.8,

$$\int_{S^{n-1}} \widehat{h_K}(\theta)\widehat{f_L}(\theta)d\theta = (2\pi)^n \int_{S^{n-1}} h_K(\theta)f_L(\theta)d\theta = n(2\pi)^n V_1(L,K),$$

so

$$V_1(L,K) < \mathrm{Vol}_n(K).$$

As in the previous lemma, this implies

$$\mathrm{Vol}_n(K) > \mathrm{Vol}_n(L).$$

□

Translated into the language of geometry (use Theorem 8.6), the result of Theorem 8.9 means that if L is a projection body, then the answer to the question of Shephard's problem is affirmative for every K. On the other hand, if K is not a projection body, then one can construct a counterexample by perturbing K. These connections with projection bodies were proved by Schneider [Sch1] in his original solution to Shephard's problem.

THEOREM 8.12. *The Shephard problem has an affirmative answer in* $\mathbb{R}^2$ *and a negative answer in* $\mathbb{R}^n$, $n \ge 3$.

PROOF. Let $n = 2$. This case is trivial, but let us give a proof that corresponds to the affirmative part of the solution to the Busemann-Petty problem. In view of Theorem 8.9, it is enough to show that for any convex symmetric body L in $\mathbb{R}^2$ such that h_L is infinitely smooth, we have $\widehat{h_L} \le 0$ on the sphere. In fact, by formula (8.10) with $n = k = 2$, for every $\xi \in S^{n-1}$,

$$\widehat{h_L}(\xi) = \pi A''_{L^*,\xi}(0) \le 0,$$

where the last inequality follows from Brunn's theorem, Theorem 2.3.

Let $n \geq 3$. To apply Theorem 8.9, we need to give an example of a symmetric convex body K so that h_K is infinitely differentiable, positive on an open subset of S^{n-1} and $f_K > 0$ on S^{n-1}. Let K' be the unit ball of $\ell^n_{4/3}$. Then the function $h_{K'}(x) = \|x\|_{\ell^n_4}$ is infinitely smooth and $\widehat{h_{K'}}$ is a continuous function on the sphere (by Corollary 3.17). This function is positive on some open subset of S^{n-1}, since otherwise, by Theorem 6.10, the space ℓ^n_4 embeds isometrically in L_1, which contradicts the solution of Schoenberg's problem, Corollary 6.12.

Define a body K by $h_K = h_{K'} + \varepsilon h_{B_2^n}$, where $\varepsilon > 0$ is such that $\widehat{h_K}$ is still positive on an open subset of S^{n-1}. Then $f_K(\theta) > 0$, $\forall \theta \in S^{n-1}$, and the result follows from Theorem 8.9. □

8.5. Remarks and further results

The main tool of the Fourier approach to projections, Theorem 8.1, was proved in [KRZ, Th. 2].

Extremal hyperplane projections of ℓ_p-balls were studied by Barthe and Naor [BN] who proved projection analogs of all corresponding results for sections of the same balls. The results for sections and projections are always dual to each other. First, Barthe and Naor proved an analog of the result of Meyer and Pajor mentioned in Section 7.3: if $1 \leq p \leq q \leq \infty$ and H is a hyperplane in $\mathbb{R}^n$, then

$$\frac{\mathrm{Vol}_{n-1}(B_p^n|H)}{\mathrm{Vol}_{n-1}(B_p^{n-1})} \leq \frac{\mathrm{Vol}_{n-1}(B_q^n|H)}{\mathrm{Vol}_{n-1}(B_q^{n-1})},$$

where $B_p^n|H$ is the projection of B_p^n to H. Putting $p = 2$, we immediately get that for $q \geq 2$, the minimal hyperplane projection of B_q^n is to the hyperplane $(1, 0, ..., 0)^\perp$, and putting $q = 2$, we see that the same projection is maximal for B_p^n with $p \leq 2$. The case of maximal projections of B_p^n with $p \geq 2$ is resolved in [BN] using formula (8.3) and then an argument similar to that in Theorem 7.7. The proof of formula (8.3) in [BN] is probabilistic. We showed in Section 8.2 that this formula follows from Theorem 8.1 in the same way that the formula of Meyer and Pajor for sections follows from Theorem 3.8 (this argument is taken from [KRZ]). Finally, the case of minimal hyperplane projections of B_p^n with $p < 2$ remains open, like the case of maximal hyperplane sections of B_p^n with $p > 2$. Barthe and Naor give a lower bound in this case, which is sharp for $p = 1$, and then present an argument showing that the answer must be as complicated as in the corresponding case for sections; see Section 7.3.

The concept of a projection body is one of the most important in convex geometry. There are many results and open problems. We again refer the reader to the books by Schneider [Sch3] and Gardner [Ga3] for details and references. The observation that a body K is a projection body if and only if its polar K^* is the unit ball of a subspace of L_1 is due to Bolker [Bl]. The Fourier analytic characterization of Theorem 8.6 was proved in [KRZ]. However, it is a result going back to P. Levy that a normed space embeds in L_1 if and only if the norm is a negative definite function.

Shephard's problem was posed in [Sh] in 1964 and represents the projection counterpart of the Busemann-Petty problem. However, unlike the Busemann-Petty problem, Shephard's problem was solved soon after it was asked, by Petty [Pe] and Schneider [Sch1], independently. Schneider found a connection between projection bodies and Shephard's problem; see Theorem 8.9 and a comment after this theorem. Lutwak's connection between intersection bodies and the Busemann-Petty problem is the section analog of Schneider's result. The Fourier analytic solution to Shephard's problem from Section 8.4 was found in [KRZ] and is similar to the solution of the Busemann-Petty problem from Section 5.2. Further results related to Shephard's problem were obtained by Ball [Ba4], who proved that it is necessary to multiply $\mathrm{Vol}_n(L)$ by $\sqrt{n}$ to make the answer affirmative in all dimensions, and by Goodey and Zhang [GZ], who solved a generalization of Shephard's problem with lower dimensional projections. Ryabogin and Zvavitch [RyZ] solved an analog of Shephard's problem for the Firey projections.

Bibliography

[Al] S. Alesker, *The α-cosine transform and intertwining integrals on real Grassmanians*, preprint.

[AB] S. Alesker and J. Bernstein, *Range characterization of the cosine transform on higher Grassmannians*, Adv. Math. 184 (2004), 367–379.

[AMM] I. Aharoni, B. Maurey and B. Mityagin, *Uniform embeddings of metric spaces and of Banach spaces into Hilbert spaces*, Israel J. Math. **52** (1985), 251–265.

[Am] D. Amir, *Isometric characterizations of inner product spaces*, Birkhäuser, Basel, 1986.

[As] R. Askey, *Radial characteristic functions*, Tech. Report no. 1262, University of Wisconsin, Madison, 1973.

[Ba1] K. Ball, *Cube slicing in* $\mathbb{R}^n$, Proc. Amer. Math. Soc. **97** (1986), 465–473.

[Ba2] K. Ball, *Some remarks on the geometry of convex sets*, Geometric aspects of functional analysis (1986/87), Lecture Notes in Math. **1317**, Springer-Verlag, Berlin-Heidelberg-New York, 1988, 224–231.

[Ba3] K. Ball, *Volumes of sections of cubes and related problems*, Geometric aspects of functional analysis (1987–88), Lecture Notes in Math. **1376**, Springer-Verlag, Berlin-Heidelberg-New York, 1989, 251–260.

[Ba4] K. Ball, *Shadows of convex bodies*, Trans. Amer. Math. Soc. **327** (1991), 891-901.

[Ba5] K. Ball, *An elementary introduction to modern convex geometry*, Flavors of geometry, Math. Sci. Res. Inst. Publ., **31**, Cambridge Univ. Press, Cambridge, 1997, 1–58.

[Ba6] K. Ball, *Convex geometry and functional analysis*, Handbook of the geometry of Banach spaces, Vol. I, North-Holland, Amsterdam, 2001, 161–194.

[Ba7] K. Ball, *Logarithmically concave functions and sections of convex sets in* R^n, Studia Math. **88** (1988), 69–84.

[Ba8] K. Ball, *Volume ratios and a reverse isoperimetric inequality*, J. London Math. Soc. (2) **44** (1991), 351–359.

[Bar] F. Barthe, *Measures unimodales et sections des boules* B_p^n, C.R. Acad. Sci. Paris **321** (1995), 865–868.

[BFM] F. Barthe, M. Fradelizi and B. Maurey, *A short solution to the Busemann-Petty problem*, Positivity **3** (1999), 95–100.

[BK] F. Barthe and A. Koldobsky, *Extremal slabs in the cube and the Laplace transform*, Adv. Math. **174** (2003), 89–114.

[BN] F. Barthe and A. Naor, *Hyperplane projections of the unit ball of* l_p^n, Discrete Comput. Geom. **27** (2002), 215–226.

[BL] Y. Benyamini and J. Lindenstrauss, *Geometric nonlinear functional analysis. Vol. 1*, American Mathematical Society Colloquium Publications, **48**, American Mathematical Society, Providence, RI, 2000.

[BR] C. Berg and P. Ressel, *Une forme abstraite du theoreme de Schoenberg*, Archiv Math. **30** (1978), 55–61.

[Bl] E. D. Bolker, *A class of convex bodies*, Trans. Amer. Math. Soc. **145** (1969), 323–345.

[Bo1] J. Bourgain, *On the Busemann-Petty problem for perturbations of the ball*, Geom. Funct. Anal. **1** (1991), 1–13.

[Bo2] J. Bourgain, *On the distribution of polynomials on high-dimensional convex sets*, Geometric aspects of functional analysis (1989–90), Lecture Notes in Math. **1469**, Springer-Verlag, Berlin-Heidelberg-New York, 1991, 127–137.

[Bo3] J. Bourgain, *On high dimensional maximal functions associated to convex bodies*, Amer. J. Math. **108** (1986), 1467–1476.

[BZ] J. Bourgain and Gaoyong Zhang, *On a generalization of the Busemann-Petty problem*, Convex geometric analysis (Berkeley, CA, 1996), 65–76, Math. Sci. Res. Inst. Publ., **34**, Cambridge Univ. Press, Cambridge, 1999.

[BDK] J. Bretagnolle, D. Dacunha-Castelle and J. L. Krivine, *Lois stables et espaces L_p*, Ann. Inst. H. Poincaré Probab. Statist. **2** (1966), 231–259.

[Bu1] H. Busemann, *A theorem on convex bodies of the Brunn-Minkowski type*, Proc. Nat. Acad. Sci. U.S.A. **35** (1949), 27–31.

[Bu2] H. Busemann, *Volumes and areas of cross-sections*, Amer. Math. Monthly **67** (1960), 248–250.

[BP] H. Busemann and C. M. Petty, *Problems on convex bodies*, Math. Scand. **4** (1956), 88–94.

[Cae] A. M. Caetano, *Weyl numbers in sequence spaces and sections of unit balls*, J. Funct. Anal. **106** (1992), 1–17.

[CKS] S. Cambanis, R. Keener, and G. Simons, *On α-symmetric multivariate distributions*, J. Multivariate Analysis **13** (1983), 213–233.

[Ca1] S. Campi, *Convex intersection bodies in three and four dimensions*, Mathematika **46** (1999), 15–27.

[Ca2] S. Campi, *Stability estimates for star bodies in terms of their intersection bodies*, Mathematika **45** (1998), 287–303.

[CR] J. P. R. Christensen and P. Ressel, *Norm dependent positive definite functions on B-spaces*, Probability in Banach spaces, IV (Oberwolfach, 1982), Lecture Notes in Math. **990**, Springer-Verlag, Berlin-Heidelberg-New York, 1983, 47–53.

[DK] S. Dilworth and A. Koldobsky, *The Fourier transform of order statistics with applications to Lorentz spaces*, Israel J. Math. **92** (1995), 411–425.

[Do] L. Dor, *Potentials and isometric embeddings in L_1*, Israel J. Math. **24** (1976), 260–268.

[DS] N. Dunford and J. T. Schwartz, *Linear operators, Part 1: General theory*, Interscience, New York, 1958.

[Ea] M. Eaton, *On the projections of isotropic distributions*, Ann. Stat. **9** (1981), 391–400.

[FGW] H. Fallert, P. Goodey and W. Weil, *Spherical projections and centrally symmetric sets*, Adv. Math. **129** (1997), 301–322.

[F] W. Feller, *An introduction to probability theory and its applications*, Wiley & Sons, New York, 1971.

[Fe] T. S. Ferguson, *A representation of the symmetric bivariate Cauchy distributions*, Ann. Math. Stat. **33** (1962), 1256–1266.

[Ga1] R. J. Gardner, *Intersection bodies and the Busemann-Petty problem*, Trans. Amer. Math. Soc. **342** (1994), 435–445.

[Ga2] R. J. Gardner, *A positive answer to the Busemann-Petty problem in three dimensions*, Annals of Math. **140** (1994), 435–447.

[Ga3] R. J. Gardner, *Geometric tomography*, Cambridge University Press, New York, 1995.

[Ga4] R. J. Gardner, *The Brunn-Minkowski inequality*, Bull. Amer. Math. Soc. (N.S.) **39** (2002), 355–405.

[GKS] R. J. Gardner, A. Koldobsky and Th. Schlumprecht, *An analytic solution to the Busemann-Petty problem on sections of convex bodies*, Annals of Math. **149** (1999), 691–703.

[GS] I. M. Gelfand and G. E. Shilov, *Generalized functions, vol. 1. Properties and operations*, Academic Press, New York, 1964.

[GV] I. M. Gelfand and N. Ya. Vilenkin, *Generalized functions, vol. 4. Applications of harmonic analysis*, Academic Press, New York, 1964.

[Gi] A. Giannopoulos, *A note on a problem of H. Busemann and C. M. Petty concerning sections of symmetric convex bodies*, Mathematika **37** (1990), 239–244.

[GiM] A. Giannopoulos and V. Milman, *Euclidean structure in finite dimensional normed spaces*, Handbook of the geometry of Banach spaces, Vol. I, 707–779, North-Holland, Amsterdam, 2001.

[Gn1] T. Gneiting, *On α-symmetric multivariate characteristic functions*, J. Multivariate Anal. **64** (1998), 131–147.

[Gn2] T. Gneiting, *Criteria of Polya type for radial positive definite functions*, Proc. Amer. Math. Soc. **129** (2001), 2309–2318.

[GLW] P. R. Goodey, E. Lutwak and W. Weil, *Functional analytic characterization of classes of convex bodies*, Math. Z. **222** (1996), 363–381.

[GW1] P. Goodey and W. Weil, *Centrally symmetric convex bodies and Radon transforms on higher order Grassmanians*, Mathematika **38** (1991), 117–133.

[GW2] P. Goodey and W. Weil, *Intersection bodies and ellipsoids*, Mathematika **42** (1995), 295–304.

[GZ] P. Goodey and Gaoyong Zhang, *Inequalities between projection functions of convex bodies*, Amer. J. Math. **120** (1998), 345-367.

[Gri] E. Grinberg, *Radon transforms on higher rank Grassmanins*, J. Diff. Geom. **24** (1986), 53–68.

[GrZ] E. Grinberg and Gaoyong Zhang, *Convolutions, transforms, and convex bodies*, Proc. London Math. Soc. (3) **78** (1999), 77–115.

[Gr] H. Groemer, *Geometric applications of Fourier series and spherical harmonics*, Cambridge University Press, New York, 1996.

[GrM] R. Grzaslewicz and J. Misiewicz, *Isometric embeddings of subspaces of L_α-spaces and maximal representation for symmetric stable processes*, Functional Analysis, Proceedings of the 1st International Workshop in Trier, 1994, 179–182.

[Ha1] H. Hadwiger, *Gitterperiodische punktmengen und isoperimetrie*, Monatsh. Math. **76** (1972), 410–418.

[Ha2] H. Hadwiger, *Radialpotenzintegrale zentralsymmetrischer rotationkörper und ungleichheitsaussagen Busemannscher art*, Math. Scand. **23** (1968), 193-200.

[HO] H. Hadwiger and D. Ohmann, *Brunn-Minkowskischer satz und isoperimetrie*, Math. Z. **66** (1956), 1–8.

[H1] S. Helgason, *The Radon transform*, Second edition, Progress in Mathematics, 5, Birkhäuser Boston, Inc., Boston, MA, 1999.

[H2] S. Helgason, *Groups and geometric analysis*, Academic Press, New York, 1984.

[He1] D. Hensley, *Slicing the cube in $\mathbb{R}^n$ and probability (bounds for the measure of a central cube slice in R^n by probability methods)*, Proc. Amer. Math. Soc. **73** (1979), 95–100.

[He2] D. Hensley, *Slicing convex bodies—bounds for slice area in terms of the body's covariance*, Proc. Amer. Math. Soc. **79** (1980), 619–625.

[Her] C. Herz, *A class of negative definite functions*, Proc. Amer. Math. Soc. **14** (1963), 670–676.

[J] F. John, *Extremum problems with inequalities as subsidiary conditions*, In Studies and Essays Presented to R. Courant on his 60th Birthday, Interscience, New York, 1948, 187–204.

[Ka] N. J. Kalton, *Banach spaces embedding into L_0*, Israel J. Math **52** (1985), 305–319.

[KK1] N. J. Kalton and A. Koldobsky, *Banach spaces embedding isometrically in L_p when $0 < p < 1$*, Proc. Amer. Math. Soc. **132** (2004), 67–76.

[KK2] N. J. Kalton and A. Koldobsky, *Intersection bodies and L_p-spaces*, Adv. Math., to appear.

[KKYY] N. J. Kalton, A. Koldobsky, V. Yaskin and M. Yaskina, *The geometry of L_0*, preprint.

[KPR] N. J. Kalton, N. T. Peck and J. W. Roberts, *An F-space sampler*, Cambridge University Press, New York, 1984.

[Kla] B. Klartag, *An isomorphic version of the slicing problem*, J. Funct. Anal. **218** (2005), 372–394.

[K1] A. Koldobsky, *The Schoenberg problem on positive-definite functions*, Algebra i Analiz **3** (1991), 78–85; translation in St. Petersburg Math. J. **3** (1992), 563–570.

[K2] A. Koldobsky, *Generalized Levy representation of norms and isometric embeddings into L_p-spaces*, Ann. Inst. H. Poincare ser. B **28** (1992), 335–353.

[K3] A. Koldobsky, *Common subspaces of L_p-spaces*, Proc. Amer. Math. Soc. **122** (1992), 207–212.

[K4] A. Koldobsky, *Characterization of measures by potentials*, J. Theor. Prob. **7** (1994), 135–145.

[K5] A. Koldobsky, *A Banach subspace of $L_{1/2}$ which does not embed in L_1 (isometric version)*, Proc. Amer. Math. Soc. **124** (1996), 155–160.

[K6] A. Koldobsky, *Inverse formula for the Blaschke-Levy representation*, Houston J. Math. **23** (1997), 95–108.

[K7] A. Koldobsky, *An application of the Fourier transform to sections of star bodies*, Israel J. Math. **106** (1998), 157–164.

[K8] A. Koldobsky, *Intersection bodies in $\mathbb{R}^4$*, Adv. Math. **136** (1998), 1–14.

[K9] A. Koldobsky, *Second derivative test for intersection bodies*, Adv. Math. **136** (1998), 15–25.

[K10] A. Koldobsky, *Intersection bodies, positive definite distributions and the Busemann-Petty problem*, Amer. J. Math. **120** (1998), 827–840.

[K11] A. Koldobsky, *A correlation inequality for stable random vectors*, Advances in stochastic inequalities (Atlanta, GA, 1997), Contemp. Math. **234**, Amer. Math. Soc., Providence, RI, 1999, 121–124.

[K12] A. Koldobsky, *Positive definite distributions and subspaces of L_{-p} with applications to stable processes*, Canad. Math. Bull. **42** (1999), 344–353.

[K13] A. Koldobsky, *A generalization of the Busemann-Petty problem on sections of convex bodies*, Israel J. Math. **110** (1999), 75–91.

[K14] A. Koldobsky, *A functional analytic approach to intersection bodies*, Geom. Funct. Anal. **10** (2000), 1507–1526.

[K15] A. Koldobsky, *On the derivatives of x-ray functions*, Arch. Math. **79** (2002), 216–222.

[K16] A. Koldobsky, *The Busemann-Petty problem via spherical harmonics*, Adv. Math. **177** (2003), 105–114.

[K17] A. Koldobsky, *Comparison of volumes by means of areas of central sections*, Adv. Appl. Math. **33** (2004), 728–732.

[K18] A. Koldobsky, *Sections of star bodies and the Fourier transform*, Harmonic Analysis at Mount Holyoke, Proceedings of AMS-IMS-SIAM Joint Summer Research Conference on Harmonic Analysis, Contemp. Math. **320**, Amer. Math. Soc., Providence, RI, 2003, 225–248.

[K19] A. Koldobsky, *Isometries of $L_p(X; L_q)$ and equimeasurability*, Indiana Univ. Math. J. **40** (1991), 677–705.

[KKo] A. Koldobsky and H. König, *Aspects of the isometric theory of Banach spaces*, Handbook of the geometry of Banach spaces, Vol. I, North-Holland, Amsterdam, 2001, 899–939.

[KL] A. Koldobsky and Y. Lonke, *A short proof of Schoenberg's conjecture on positive definite functions*, Bull. London Math. Soc. **31** (1999), 693–699.

[KRZ] A. Koldobsky, D. Ryabogin and A. Zvavitch, *Projections of convex bodies and the Fourier transform*, Israel J. Math. **139** (2004), 361–380.

[KYY] A. Koldobsky, V. Yaskin and M. Yaskina, *Modified Busemann-Petty problem on sections of convex bodies*, preprint.

[KZ] A. Koldobsky and M. Zymonopoulou, *Extremal sections of complex l_p-balls, $0 < p \leq 2$*, Studia Math. **159** (2003), 185–194.

[Kom] J. Komlos, *A generalization of a problem of Steinhaus, Acta Math. Acad. Sci. Hungar.*, **18** (1967), 217-229.

[Ko] H. König, *Isometric embeddings of Euclidean spaces into finite dimensional ℓ_p-spaces*, Banach Center Publ. **34** (1995), 79–87.

[Kue] J. Kuelbs, *Positive definite symmetric functions on linear spaces*, J. Math. Anal. Appl. **42** (1973), 413–426.

[Kur] Yu. G. Kuritsyn, *Multidimensional versions and two problems of Schoenberg*, Problems of Stability of Stochastic Models, VNIISI, Moscow, 1989, 72–79.

[KuS] Yu. G. Kuritsyn and A. V. Shestakov, *On α-symmetric distributions*, Theory Probab. Appl. **29** (1984), 769–771.

[Kw] S. Kwapien, *Problem 3*, Studia Math. **38** (1970), 469.

[La] P. S. Laplace, *Théorie analytique des probabilités*, Paris, 1812.

[Lv] B. Ya. Levin, *Lectures on entire functions*, Amer. Math. Soc., Providence, RI, 1996.

[Le] P. Lévy, *Théorie de l'addition de variable aléatoires*, Gauthier-Villars, Paris, 1937.

[LR] D. G. Larman and C. A. Rogers, *The existence of a centrally symmetric convex body with central sections that are unexpectedly small*, Mathematika **22** (1975), 164–175.

[Li] J. Lindenstrauss, *On the extension of operators with finite dimensional range*, Illinois J. Math. **8** (1964), 488–499.

[LT] J. Lindenstrauss and L. Tzafriri, *Classical Banach spaces*, Lecture Notes in Math. **338**, Springer-Verlag, Berlin-Heidelberg-New York, 1973.

[Lin] W. Linde, *Probability in Banach spaces–stable and infinitely divisible distributions*, Second edition, Wiley-Interscience, Chichester, 1986.

[Lis] A. Lisitsky, *One more proof of Schoenberg's conjecture*, unpublished manuscript, 1991.

[Lo] Y. Lonke, *On zonoids whose polars are zonoids*, Israel J. Math. **102** (1997), 1–12.

[Luc] E. Lukacs, *Characteristic functions*, Griffin, London, 1970, second edition.

[Lus] L. A. Lusternik, *Die Brunn-Minkowskischer Ungleichung fur beliebige messbare Mengen*, Dokl. Acad. Sci. SSSR **8** (1935), 55–58.

[Lu1] E. Lutwak, *Intersection bodies and dual mixed volumes*, Adv. Math. **71** (1988), 232–261.

[Lu2] E. Lutwak, *The Brunn-Minkowski-Firey theory. I. Mixed volumes and the Minkowski problem*, J. Diff. Geom. **38** (1993), 131–150.

[Lu3] E. Lutwak, *The Brunn-Minkowski-Firey theory. II. Affine and geominimal surface areas*, Adv. Math. **118** (1996), 244–294.

[LYZ1] E. Lutwak, D. Yang and Gaoyong Zhang, *A new ellipsoid associated with convex bodies*, Duke Math. J. **104** (2000), 375–390.

[LYZ2] E. Lutwak, D. Yang and Gaoyong Zhang, *L_p-affine isoperimetric inequalities*, J. Diff. Geom. **56** (2000), 111–132.

[LYZ3] E. Lutwak, D. Yang and Gaoyong Zhang, *On the L_p-Minkowski problem*, Trans. Amer. Math. Soc. **356** (2004), 4359–4370.

[LYZ4] E. Lutwak, D. Yang and Gaoyong Zhang, *Volume inequalities for subspaces of L_p*, J. Diff. Geom., to appear.

[LV] Y. Lyubich and L. Vaserstein, *Isometric embeddings between classical Banach spaces*, Geom. Dedicata **47** (1993), 327–362.

[Ma] B. Maurey, *Théoremes de factorization pour les operatèurs linéaires à valeurs dans les espaces L_p*. Astérisque **11** (1974), Société Math. de France.

[Me] M. Meyer, *On a problem of Busemann and Petty*, unpublished manuscript.

[MeP] M. Meyer and A. Pajor, *Sections of the unit ball of ℓ_q^n*, J. Funct. Anal. **80** (1988), 109–123.

[Mi] E. Milman, private communication.

[MiP] V. Milman and A. Pajor, *Isotropic position and inertia ellipsoids and zonoids of the unit ball of a normed n-dimensional space*, Geometric aspects of functional analysis (1987–88), Lecture Notes in Math. **1376**, Springer-Verlag, Berlin-Heidelberg-New York, 1989, 64–104.

[MiS] V. Milman and G. Schechtman, *Asymptotic theory of finite-dimensional normed spaces*, Lecture Notes in Math. **1200**, Springer-Verlag, Berlin-Heidelberg-New York, 1986.

[Mis1] J. Misiewicz, *On norm dependent positive definite functions*, Bull. Acad. Sci. Georgian SSR **130** (1988), 253–256.

[Mis2] J. Misiewicz, *Positive definite functions on ℓ_∞*, Stat. Probab. Let. **8** (1989), 255–260.

[Mis3] J. Misiewicz, *Substable and pseudo-isotropic processes—connections with the geometry of subspaces of L_α-spaces*, Dissertationes Math. (Rozprawy Mat.) **358** (1996).

[MisR] J. Misiewicz and Cz. Ryll-Nardzewski, *Norm dependent positive definite functions*, Lecture Notes in Math. **1391**, Springer-Verlag, Berlin-Heidelberg-New York, 1987, 284–292.

[MisS] J. Misiewicz and C. L. Scheffer, *Pseudo isotropic measures*, Nieuw Arch. Wisk. **8** (1990), 111–152.

[Mu] C. Müller, *Spherical harmonics*, Springer-Verlag, Berlin-Heidelberg-New York, 1966.

[NP] F. L. Nazarov and A. N. Podkorytov, *Ball, Haagerup, and distribution functions*, Complex analysis, operators, and related topics, Oper. Theory Adv. Appl. **113**, Birkhäuser, Basel, 2000, 247–267.

[Ni] E. M. Nikishin, *Resonance theorems and superlinear operators,* Uspeki Mat. Nauk. **25** (1970), 129-191.

[O] K. Oleszkiewicz, *On p-pseudostable random variables, Rosenthal spaces and l_p^n ball slicing*, Geometric aspects of functional analysis, Lecture Notes in Math. **1807**, Springer-Verlag, Berlin-Heidelberg-New York, 2003, 188–210.

[OP] K. Oleszkiewicz and A. Pelczynski, *Polydisc slicing in C^n*, Studia Math. **142** (2000), 281–294.

[Pa] M. Papadimitrakis, *On the Busemann-Petty problem about convex, centrally symmetric bodies in $\mathbb{R}^n$*, Mathematika **39** (1992), 258–266.

[Pe] C. M. Petty, *Projection bodies*, Proc. Coll. Convexity (Copenhagen 1965), Kobenhavns Univ. Mat. Inst., 234-241.

[Pi1] G. Pisier, *Factorization of operators through $L_{p,\infty}$ or $L_{p,1}$ and non-commutative generalizations*, Math. Ann. **276** (1986), 105-136.

[Pi2] G. Pisier, *The volume of convex bodies and Banach space geometry,* Cambridge Tracts in Mathematics, **94**, Cambridge University Press, Cambridge, 1989.

[Po1] G. Polya, *Berechnung eines bestimmten integrals*, Math. Ann. **74** (1913), 204–212.

[Po2] G. Polya, *On the zeroes of an integral function represented by Fourier's integral*, Messenger Math. **52** (1923), 185–188.

[PS] G. Polya and G. Szego, *Aufgaben und lehrsatze aus der analysis*, Springer-Verlag, Berlin-New York, 1964.

[Ri] D. St. P. Richards, *Positive definite symmetric functions on finite dimensional spaces. 1. Applications of the Radon transform*, J. Multivariate Analysis **19** (1986), 280–298.

[Rb1] B. Rubin, *Inversion formulas for the spherical Radon transform and the generalized cosine transform*, Adv. Appl. Math. **29** (2002), 471–497.

[Rb2] B. Rubin, *Notes on Radon transforms in integral geometry*, Fract. Calc. Appl. Anal. **6** (2003), 25–72.

[RZ] B. Rubin and Gaoyong Zhang, *Generalizations of the Busemann-Petty problem for sections of convex bodies*, J. Funct. Anal. **213** (2004), 473–501.

[Ru] W. Rudin, *Functional analysis*, McGraw-Hill, New York, 1973.

[RyZ] D. Ryabogin and A. Zvavitch, *Fourier transform and Firey projections of convex bodies*, Indiana Univ. Math. J. **53** (2004), 667–682.

[Sch1] R. Schneider, *Zu einem problem von Shephard über die projektionen konvexer Körper*, Math. Z. **101** (1967), 71-82.

[Sch2] R. Schneider, *Zonoids whose polars are zonoids,* Proc. Amer. Math. Soc. **50** (1975), 365-368.

[Sch3] R. Schneider, *Convex bodies: the Brunn-Minkowski theory*, Cambridge University Press, Cambridge, 1993.

[SW] R. Schneider and W. Weil, *Integralgeometrie*, B.C.Teubner, Stuttgart, 1992.

[Sc1] I. J. Schoenberg, *Metric spaces and positive definite functions*, Trans. Amer. Math. Soc. **44** (1938), 522–536.

[Sc2] I. J. Schoenberg, *Metric spaces and completely monotone functions*, Annals of Math. **39** (1938), 811–841.

[Se] V. I. Semyanistyi, *Some integral transformations and integral geometry in an elliptic space*, Trudy Sem. Vector. Tenzor. Anal. **12** (1963), 397-411.

[Sh] G. C. Shephard, *Shadow systems of convex bodies*, Israel J. Math. **2** (1964), 229-306.

[Sz] S. J. Szarek, *On the best constants in the Khinchine inequality*, Studia Math. **58** (1976), 197–208.

[Th] J. A. Thorpe, *Elementary topics in differential geometry*, Undergraduate Texts in Mathematics, Springer-Verlag, Berlin-Heidelberg-New York, 1979.

[Va] J. D. Vaaler, *A geometric inequality with applications to linear forms*, Pacific J. Math. **83** (1979), 543–553.

[Wi] D. V. Widder, *The Laplace transform*, Princeton University Press, Princeton, NJ, 1941.

[W] P. Wojtaszczyk, *Banach spaces for analysts*, Cambridge Studies in Advanced Mathematics, 25. Cambridge University Press, Cambridge, 1991.

[Y1] V. Yaskin, *The Busemann-Petty problem in spherical and hyperbolic spaces*, preprint.

[Y2] V. Yaskin, *A solution to the lower dimensional Busemann-Petty problem in the hyperbolic space*, preprint.

[Za1] V. Zastavnyi, *Positive definite norm dependent functions*, Dokl. Russian Acad. Nauk. **325** (1992), 901–903.

[Za2] V. Zastavnyi, *Positive definite functions depending on the norm*, Russian J. Math. Phys. **1** (1993), 511–522.

[Za3] V. Zastavnyi, *On positive definiteness of some functions*, J. Multivariate Anal. **73** (2000), 55–81.

[Zh1] Gaoyong Zhang, *Centered bodies and dual mixed volumes*, Trans. Amer. Math. Soc. **345** (1994), 777–801.

[Zh2] Gaoyong Zhang, *Intersection bodies and Busemann-Petty inequalities in* $\mathbb{R}^4$, Annals of Math. **140** (1994), 331–346.

[Zh3] Gaoyong Zhang, *Sections of convex bodies*, Amer. J. Math. **118** (1996), 319–340.

[Zh4] Gaoyong Zhang, *A positive answer to the Busemann-Petty problem in four dimensions*, Annals of Math. **149** (1999), 535–543.

[Zh5] Gaoyong Zhang, *Intersection bodies and polytopes*, Mathematika **46** (1999), 29–34.

[Zo] V. M. Zolotarev, *One-dimensional stable distributions*, Amer. Math. Soc., Providence, 1986.

[Zv1] A. Zvavitch, *Gaussian measure of sections of convex bodies*, Adv. Math. **188** (2004), 124-136.

[Zv2] A. Zvavitch, *The Busemann-Petty problem for arbitrary measures*, Math. Ann., to appear.

Index